The Jackson Elk Herd

THE JACKSON ELK HERD

Intensive wildlife management in North America

Mark S. Boyce
Department of Zoology and Physiology
University of Wyoming, Laramie, Wyoming

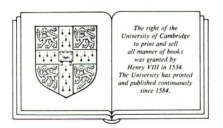

The right of the
University of Cambridge
to print and sell
all manner of books
was granted by
Henry VIII in 1534.
The University has printed
and published continuously
since 1584.

CAMBRIDGE UNIVERSITY PRESS

Cambridge

New York   Port Chester

Melbourne   Sydney

Published by the Press Syndicate of the University of Cambridge
The Pitt Building, Trumpington Street, Cambridge CB2 1RP
40 West 20th Street, New York, NY 10011, USA
10 Stamford Road, Oakleigh, Melbourne 3166, Australia

First published 1989

Printed in Great Britain at the University Press, Cambridge

British Library cataloguing in publication data

Boyce, Mark S.
The Jackson elk herd : intensive wildlife
management in North America.
1. Wyoming. Jackson Hole. Elk. Management
I. Title
639.9'797357

Library of Congress cataloguing in publication data

Boyce, Mark S.
The Jackson elk herd.
Bibliography: p.
Includes index.
1. Elk – Wyoming – Jackson Hole Region.   2. Wildlife
management – Wyoming – Jackson Hole Region.   3. Mammals –
Wyoming – Jackson Hole Region.   I. Title.
QL737.U55B63   1989   639.9'797357   88-28530

ISBN 0 521 34147 7

CONTENTS

For Evie and Cody

PREFACE

In 1912, Theodore Roosevelt warned that feeding hay to elk was a dangerous prospect for elk in northwestern Wyoming, and could only end in disaster. Visions of overpopulation, habitat destruction, severe winter mortality, disease and chaos as a consequence of supplemental feeding have been painted by wildlife biologists ever since. Yet, after 75 years of winter feeding, elk in Jackson Hole are abundant and productive, offering consumptive and nonconsumptive use opportunities for millions of visitors to the valley every year.

I could easily promote the Jackson elk herd as one of wildlife management's great success stories. Yet, there is still too much controversy surrounding the elk management program for my case to be credible. Much of this controversy ultimately stems from abundant misinformation, as well as the complexity of elk ecology and management in Jackson Hole: complexity generated by a migrating elk herd with ranges overlapping the jurisdiction of several management agencies and private landowners; complexity reinforced by federal legislation which permits hunting for elk within the boundaries of Grand Teton National Park. Despite the difficulties and headaches that these complexities have caused resource managers, I am fascinated by the way it all fits together. And it all happens in one of the most spectacular valleys in the world!

Since establishment of the National Elk Refuge in 1912, agency biologists and other personnel have collected various data on the Jackson elk herd. In this report I evaluate and synthesize these data, with particular focus on the period since the establishment of the Jackson Hole Cooperative Elk Studies Group in 1959. Most of these data are from files of the four agencies participating in the Studies Group, or in unpublished reports of the Wyoming Game and Fish Department and the National Park Service. To facilitate future research on the Jackson elk herd, I have prepared LOTUS files for a PC microcomputer containing all of the raw data reviewed in this book. At this time, these files are maintained by the National Elk Refuge and by the Wyoming Game and Fish Department in Jackson.

For any scientific enquiry, appropriate application of statistical inference requires (1) hypothesis formulation, (2) design of an experiment to test the hypothesis, (3) collection of data, and (4) application of test statistics to accept or reject a null hypothesis. In a study such as this one, where the data have already been collected, I am constrained in the design of the experiment by the methodologies already employed in collecting the data. Therefore, I must rely upon my familiarity with the data base to extract data appropriate to test hypotheses.

The key in this analysis is to identify relevant hypotheses. Management issues stimulate relevant hypotheses which are then evaluated from the data base. Changes in management practices during the 75 years for which data exist on the Jackson herd can be viewed as experiments permitting scientific inference. As a scientific investigation, such an approach necessarily suffers from lack of replication because management requires such grand-scale manipulations. Our only hope for replication rests with corroboration among studies on other populations (e.g., Houston, 1982).

I begin in Chapter 1 with a historical review and a presentation of principal management problems for the Jackson elk herd. Following chapters address questions which must be answered to solve these problems, and my best effort to answer these questions using existing data.

Chapter 2 begins with a description of subunits of the herd and their seasonal ranges. This is followed by an evaluation of the hypothesis that the major determinants of the timing and spatial pattern of elk migration are weather and hunter kill.

Chapter 3 is an overview of factors which determine fluctuations in population size and hunter kill. In Chapter 4 I review data relevant to the ecology of the elk herd, particularly habitat relationships. This is followed by four chapters which review management by the four agencies responsible for managing the Jackson elk herd. In these chapters I discuss the implications of results from Chapters 2–4 to management of the herd, and relate how actual management compares with this. Chapter 9 summarizes the public's desires from the elk management program, and emphasizes the importance of the Jackson elk herd in the broader context of the Greater Yellowstone Ecosystem.

The final chapter summarizes major conclusions stemming from this analysis and ends with recommendations for (1) collecting adequate data to ensure sound management, (2) management alternatives and how present management might be improved, and (3) research needed to fill gaps in our understanding of the population dynamics and ecology of the Jackson elk herd.

Too frequently, perhaps, it is easy for me to judge difficult management issues, because I am protected by an ivory tower. Ultimately, it will be the resource managers in Jackson Hole that will make the decisions that really matter. I hope that my efforts will make those management decisions easier and better.

ACKNOWLEDGMENTS

This study represents the results of research supported by the Jackson Hole Cooperative Elk Studies Group. Participating agencies include: Bridger-Teton National Forest, US Forest Service, United States Department of Agriculture; Grand Teton National Park, National Park Service, United States Department of Interior; National Elk Refuge, US Fish and Wildlife Service, United States Department of Interior; and the Wyoming Game and Fish Department, State of Wyoming. Administration was coordinated by Dr Kenneth Diem, Director of the University of Wyoming – National Park Service Research Center.

I am indebted to the members of the Jackson Hole Cooperative Elk Studies Group for (1) suggesting the need for this research effort, (2) securing funding for the research, (3) collecting much of the data, (4) giving me access to their files, (5) assisting in the interpretation of data, (6) reviewing drafts of this report, and (7) unrelenting encouragement and support. The key personnel have been members of the Group's Technical Committee including Floyd Gordon from Bridger-Teton National Forest, Garvice Roby of the Wyoming Game and Fish Department, Bruce Smith of the National Elk Refuge, and Robert Wood of Grand Teton National Park. Each of these individuals has invested many hours of their already overtaxed schedules to assist me with my research. I have also enjoyed support from the Group's Advisory Council composed of Reid Jackson, recently retired Supervisor of Bridger-Teton National Forest; Jack Stark, Superintendent of Grand Teton National Park; Thomas Toman, District Supervisor for the Wyoming Game and Fish Department; and John Wilbrecht, Manager of the National Elk Refuge. Special thanks are due to Bill Barmore, recently retired from Grand Teton National Park, for his assistance throughout the study.

In addition to these key individuals, many other agency personnel assisted me in various ways. From the Wyoming Game and Fish Department, I thank Rex Corsi, Doug Crowe, Walt Gasson, Harry Harju, Bill Hepworth, Marvin

Hockley, Dave Lockman, Dave Mobley, Pete Petera, Dale Strickland, Kent Schmidlin, Tom Thorne, Lee Wollrab, and Jim Yorgason. From the USDA, Forest Service, I thank Wayne Bills, Chuck Birkmeyer, Al Boss, Marc Childress, Norbert DeByle, Dave Griffel, George Gruell, David Johnson, Joe Kinsella, Gary Marple, Ernie Nunn, Carl Pence, and Gene Smalley. I thank Norm Bishop, Don Despain, Marshall Gingery, Pete Hayden, Mary Meagher, and Frank Singer from the National Park Service. And I am grateful to Jim Griffin, Rees Madsen and John Oldemeyer from the US Fish and Wildlife Service.

Many Jackson area guides, outfitters and concerned citizens discussed various aspects of elk management with me and offered helpful insights. These include Mardy Murie, Clifford Hanson, Tim Clark, Albert Feuz, Dr Ken Griggs, Joanna Johnson, Slim Lawrence, Rod Lucas, Paul Luten, Steve Minta, Steve Robertson, Glen Taylor, Ridge Taylor, Don Turner, Paul von Gontard, and Ernie Wampler. I learned much from others whose names I have forgotten or never knew.

Several people at the University of Wyoming assisted me with data analysis and offered statistical advice. These include Lyman McDonald, Don Anderson, Peter Lenth, Fred Doll, Steve Bieber and Bob Cochran. I am grateful to Deirdre Murphy and Julie Wasserburger for entering data onto my computer, Elizabeth Rahel for drafting graphs and maps, and Archie Reeve for sorting tag returns.

I enjoyed many hours of discussion on elk ecology with Tim Clutton-Brock and Morris Gosling. John Hart, Department of Botany and Plant Pathology at Michigan State University, shared his observations on elk–aspen interactions prior to publication. Douglas Woody informed me of APHIS' brucellosis programs in Wyoming. Tex Taylor from the University of Wyoming Agricultural Experiment Station provided his data on the guides and outfitters industry. Bob Bergstrom offered unpublished information on lungworm in elk. U. S. Seal and Al Franzmann helped me interpret elk blood chemistry. I thank Pinkney Wood of New Orleans, Louisiana, and Doug Houston of Olympic National Park, Washington, for valuable correspondence. I am grateful to Ludwig Carbyn, John Hart, Fred Lindzey, Cliff Martinka, Evelyn Merrill, Robin Pellew, and Tony Sinclair for reviews of the manuscript.

Finally, I am indebted to my colleagues and students at the University of Wyoming for assistance, and for providing a great atmosphere for doing research. Particular mention is due Stan Anderson, Alan Beetle, Steve Buskirk, David Duvall, Bill Gern, Rich Guenzel, Dave Gulley, Larry Irwin, Greg Johnson, Dennis Knight, Fred Lindzey, George Menkens, Evelyn Merrill, John Sauer, Larry Shults, Bill Standley and Nancy Stanton.

Chapter 1.

Historical overview and introduction to management of the Jackson elk herd

Jackson Hole, a mountain-rimmed valley of the Snake River in northwestern Wyoming, affords one of the most scenic landscapes in the world (Figure 1.1). The valley is bordered on the west by Grand Teton National Park and the spectacular Teton Range. Immediately to the north is Yellowstone National Park, the first National Park, and the largest Park in the lower 48 states of the United States.

Figure 1.1 Jackson Hole is bordered on the west by the majestic Teton Range. This view is near Taggart Lake in Grand Teton National Park. Photo by M. S. Boyce, 17 June 1986.

Until recently, Jackson Hole hosted the largest elk (*Cervus elaphus* Linnaeus 1758) herd in the world, which still ranks second only to the northern Yellowstone herd (see Houston, 1982). The Jackson elk herd numbers between 11 000 and 14 000 animals. The elk undergo a seasonal migration, probably more extensive than for any elk herd in North America, traveling up to 100 km from summer ranges to winter ranges and feedgrounds.

Figure 1.2 Bighorn sheep occupy high-elevation ranges in the Gros Ventre Mountains and a few occur in the Teton Range. A small population of bighorns winter immediately east of the National Elk Refuge, and occasionally visit the Refuge. Photo by M. S. Boyce, February 1977.

Although elk are certainly the most abundant ungulate in the area, elk share Jackson Hole with a diverse array of other wildlife. Other herbivores include bighorn sheep (*Ovis canadensis*, Figure 1.2), bison (*Bison bison*), mule deer (*Odocoileus hemionus*, Figure 1.3), moose (*Alces alces*, Figure 1.4), pronghorn (*Antilocapra americana*), plus many smaller species. The carnivore fauna includes grizzly bears (*Ursus arctos*), black bears (*Ursus americanus*), coyotes (*Canis latrans*), wolverine (*Gulo gulo*), otter (*Lutra canadensis*), red fox (*Vulpes vulpes*), lynx (*Lynx canadensis*), marten (*Martes americana*), and cougar (*Felis concolor*). Notable among the avifauna are Trumpeter Swans (*Olor buccinator*), Golden Eagles (*Aquila chrysaëtos*), Bald Eagles (*Haliaëetus leucocé-*

phalus), Blue Grouse (*Dendragapus obscurus*), Sage Grouse (*Centrocércus urophasiánus*), Sandhill Cranes (*Grus canadensis*), Whooping Cranes (*Grus americana*), Ravens (*Corvus corax*), and Mountain Bluebirds (*Siália currucoides*). Area streams and lakes are world-famous for trout fishing. Only the wolf (*Canis lupus*) is missing from this spectacular Rocky Mountain fauna.

Figure 1.3 Mule deer near Leek's Marina in Grand Teton National Park. Mule deer occur at low densities throughout the Jackson Elk Herd Unit. Deer usually include more browse in their diets than do elk. Photo by M. S. Boyce, July 1985.

The North American elk is the same species as the red deer of Europe; in New Zealand individuals introduced from North America interbreed freely with those from Europe (Caughley, 1971). Nevertheless, the elk is a much larger animal and quite different in appearance. One striking difference between red deer and elk is the mating vocalization of males. Red deer stags 'roar' with a deep sound not unlike that of a bellowing cow. In contrast, bull elk employ a high-pitched whistle seeming uncharacteristic for such a massive animal.

Despite the spectacular wilderness setting for the Jackson elk herd, overlapping two of America's grandest National Parks, it is one of the most intensively managed elk herds in North America. Winter feeding supports some 9000 to 11000 elk, with hay purchased by the federal government and the State of Wyoming. The herd is extensively harvested, supporting an average annual

Figure 1.4 Female moose east of Moran in Grand Teton National Park. Most moose in the Jackson Elk Herd Unit are dependent upon riparian willow for winter forage. Photo by M. S. Boyce, January 1977.

hunter kill of approximately 3000 animals, including a controversial annual cull of elk within the boundaries of Grand Teton National Park. Although many find the feeding and culling operations offensive for various reasons, I argue that this program is defensible.

To appreciate the intensive management of the Jackson elk herd, it is essential to understand the history of human settlement and elk management in the valley.

Historical overview

The first agricultural settlers arrived in Jackson Hole in 1884, and within the next 25 years had homesteaded the finest elk winter range in the valley. Fences were constructed, cattle were grazed, hay was cut, and as a consequence elk were displaced from winter range. Elk suffered extensive mortality during severe winters and commonly raided private haystacks (Wilbrecht & Robbins, 1979).

The tough winter of 1909 was particularly serious when an estimated 20000 to 30000 elk attempted to survive the winter in Jackson Hole. Local citizens appealed for help from their government representatives, and in 1910, $5000 was appropriated by the Wyoming Legislature for the purchase of feed. But still, many hundred elk died that winter (Sheldon, 1927).

As a consequence of considerable public support, on 17 February 1911, the Wyoming State Legislature officially requested support from the US Congress to assist with conservation and feeding of wintering elk. Congress acted more quickly in those days, and by 4 March 1911, an appropriation of $20000 was made for feeding and otherwise 'protecting' elk in Jackson Hole, and for financing transplants of elk from Jackson Hole to other areas.

The US Biological Survey evaluated the situation, and made recommendations on possible solutions (Preble, 1911). One recommendation was that a refuge for elk be established near the town of Jackson. On 10 August 1912, the National Elk Refuge was created by an act of Congress. This act appropriated $45000 for land acquisition and maintenance of operations, as well as a land allocation of 1000 acres (405 ha) from public domain.

Gradually the Refuge expanded. In 1919, US Forest Service lands adjacent to the eastern border of the Refuge were excluded from livestock grazing to increase winter forage for elk. In 1927, the Izaak Walton League, a private conservation organization, purchased 712.5 ha which they donated to the Refuge. Other federally-financed purchases of private lands have gradually increased the Refuge from 1117.4 ha in 1912 to its present size of 9833.6 ha.

When the National Elk Refuge was established, Jackson was primarily a cattle-ranching community, and it is clear from early reports that conflicts between elk and cattle ranching were the impetus behind the initiation of the supplemental feeding program. Feeding was initiated primarily to reduce depredations on cattlemen's hay supplies (Wilbrecht & Robbins, 1979). Also, feeding was necessary to sustain the elk herd because cattle ranching had usurped elk winter range.

Grand Teton National Park was not established until 1929 and was not expanded to its present size until 1950. Considerable controversy surrounded the establishment and enlargement of the Park (Righter, 1982). One of the main concerns of the State of Wyoming was the loss of management control for part of the Jackson elk herd. This was of particular concern because of the State's financial contribution to feeding elk and the possibility that elk protected in the Park would overpopulate the valley and dominate in numbers on the Refuge and state feedgrounds (Blunt, 1950). Consequently, as a compromise solution, enabling legislation for Grand Teton National Park (Public Law 81–787; 64 Stat. 849) explicitly required that reduction of elk be permitted in the Park if it was found necessary by the National Park Service and the Wyoming Game and Fish Commission.

Sport hunting in National Parks is not permitted under an International

Treaty for Nature Protection and Wildlife Preservation in the Western Hemisphere which the United States signed with 17 other countries in 1942, and therefore it appears that deputization of hunters as park rangers is a technicality to circumvent conflict with this agreement (Cole, 1969). Anyone licensed for elk hunting by the State of Wyoming may apply for a limited quota permit to participate in the elk reduction program in Grand Teton National Park, with the additional stipulation that the hunter must have taken an approved hunter safety course, usually one accredited by the National Rifle Association.

For several years after the establishment of Grand Teton National Park, the National Park Service and the Wyoming Game and Fish Department debated the interpretation of Public Law 81–787. In particular, they disagreed over the meaning of the statement that elk reduction programs would occur 'when it is found necessary for the purpose of proper management and protection of the elk', and they disagreed over just where within the Park the elk reduction program would be conducted (Murie, 1951a, 1951b, 1952, 1953). Public Law 81–787 does not clearly state criteria that must be met to make the reduction program within the Park 'necessary', and early interpretations by the National Park Service suggested that the onus was upon the Wyoming Game and Fish Department to demonstrate such necessity (Murie, 1951a, 1952). The National Park Service clearly does not approve of hunting in National Parks, and would prefer to reduce or eliminate hunting in Grand Teton National Park. Nevertheless, the Park Service has agreed to the necessity of elk reduction programs each year since 1950, except for 1959 and 1960, when the Wyoming Game and Fish Department requested that there be no reduction program.

The study area

Boundaries of the Jackson Herd Unit are defined by the Wyoming Game and Fish Department for purposes of management (Figure 1.5). Ideally, herd unit boundaries are established so that less than 10% interchange occurs between adjacent populations (Gasson, 1987: 10). The area of about 5490 km² outlined by Figure 1.5 encompasses the Jackson herd unit, excluding Yellowstone National Park. Of this, about 203 km² are water, steep mountain peaks, or the town of Jackson; leaving about 5288 km² potentially occupied by elk, at least during part of the year. In addition, summer ranges used by the Jackson elk herd in southern Yellowstone National Park total approximately 800 km². The area within the Jackson herd unit ranges in elevation from 1900 m near the town of Jackson to 4197 m at the peak of the Grand Teton. In actuality, elk do not range high into the Teton Range, and seldom use summer ranges higher than 3300 m.

Climate is cool with mean annual temperature at Jackson of 3.27 °C.

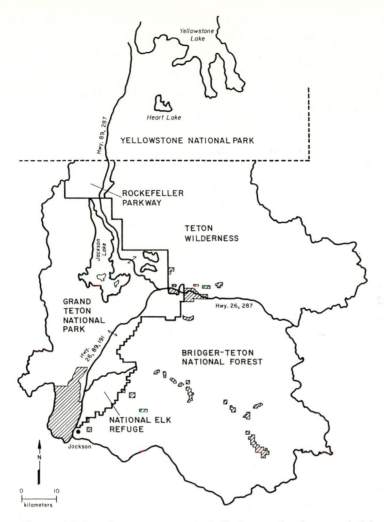

Figure 1.5 Land management jurisdiction on lands occupied by the Jackson elk herd. Cross-hatched areas are private lands. All other lands are owned by federal or state government.

Summer temperatures rarely exceed 32 °C, whereas winter temperatures below − 40 °C are not uncommon. Precipitation is heaviest in May and December, with average annual precipitation of 402 mm in Jackson. Snow accumulation is much heavier in northern portions of the herd unit (see Chapter 2). Most of the area is upland forest dominated by lodgepole pine or Douglas fir at lower elevations and Engelmann spruce-subalpine fir stands at mesic sites and higher elevations. Vegetation types and plant succession within the herd unit are characterized in more detail by Cole (1969).

Current management of the Jackson elk herd

Four agencies are primarily responsible for managing the Jackson elk herd and lands which it occupies. Figure 1.5 shows land management jurisdictions for these agencies. In Chapters 5–8 I discuss management by each of these agencies in detail.

The four agencies are (1) the Wyoming Game and Fish Department which manages hunting for the herd and contributes to the feeding program, (2) the US Forest Service which manages habitats on Bridger–Teton National Forest which constitutes more than half of area occupied by the Jackson elk herd unit, (3) the US Fish and Wildlife Service which is responsible for the National Elk Refuge where most of the elk are fed during winter, and (4) the US Park Service, in charge of Grand Teton National Park and Yellowstone National Park which provide summer range for a majority of the elk in the herd unit.

Each agency approaches its responsibility with different management philosophies and priorities. Interests of hunters play an important role in management by the Game and Fish Department, whereas priority is given to timber management by the US Forest Service. Both the Fish and Wildlife Service and the Park Service are within the US Department of Interior, but the two agencies have remarkably different philosophies and priorities for management of the Jackson elk herd. While the Park Service would prefer to avoid any intervention with the elk, the Fish and Wildlife Service manages elk intensively with range improvement on the Refuge in summer and supplemental feeding in winter.

Jackson Hole Cooperative Elk Studies Group

The charge created by Congress in Public Law 81–787 to collect and evaluate data relevant to the management of the Jackson elk herd ultimately led to the formation of the Jackson Hole Cooperative Elk Studies Group on 1 July 1958. The Studies Group consists of representatives from each of the four agencies involved in managing the herd and considers the full spectrum of activities affecting the Jackson elk herd and offers joint interagency recommendations when appropriate. The Group consists of two committees. An Advisory Council is composed of (1) the Superintendent of Grand Teton National Park, (2) the Supervisor for Bridger–Teton National Forest, (3) the Director of the Wyoming Game and Fish Department, and (4) the Manager of the National Elk Refuge. A Technical Committee, which studies issues and makes recommendations to the Advisory Council, is composed of one or two professional biologists or natural resource managers from each of the four agencies. In addition, representatives from Yellowstone National Park, Targhee National Forest, and the public are invited to attend annual meetings.

Management issues

Numerous controversies surround the management of the Jackson elk herd. Many of these are classic issues in wildlife management. Included are conflicts between agricultural interests and wildlife, controversies regarding the existence and nature of the feeding program, and conflicting priorities for management by various agencies whose land is occupied by the elk herd.

Land uses

Logging and livestock grazing are potential threats to elk habitats within the Jackson elk herd unit, in part because management plans by the US Forest Service are in a state of flux. Many areas of the National Forest have already been extensively logged, and improved road access into these areas has resulted in excessive hunter kills along certain migration routes. Current plans offer promising resolution to conflicts between elk and logging by scheduling timber exploitation to balance cover and forage requirements for elk. However, inadequate attention has been given to unique habitat requirements for migrating elk.

Livestock grazing occurs on extensive areas of elk range. The Forest Service Plan proposes an increase in cattle grazing, particularly on elk winter ranges (Bridger–Teton National Forest, 1986b: Management Area Prescription Map Appendix), even though conflicts between cattle grazing and elk have persisted on some of these areas. Fire has rarely been used to improve elk habitat *per se*, although the value of fire management to enhance elk range has been recognized for some time. Rather, most fire management employed on the National Forest is to enhance livestock range and any benefits to elk are incidental. Brucellosis in elk poses a potential management conflict because the disease can be transmitted to cattle.

Nevertheless, despite some persisting management conflicts, management decisions by the Forest Service regarding logging and cattle grazing give some consideration to elk habitats, and wildlife values are recognized more than in the past. Perhaps the most critical issues surrounding management of the Jackson elk herd are those relating to changing land use patterns. Development of private lands and persistent efforts by the oil and gas industry to exploit energy resources threaten elk winter ranges and migration corridors. Federal mining laws give high priority to mineral interests in the United States, often at the expense of other resource values.

Elk hunting in a National Park

A long-standing issue in the management of the Jackson elk herd is the herd reduction program within Grand Teton National Park. The Park's

Natural Resources Management Plan proposes an experimental reduction or closure of the Park hunt. The Wyoming Game and Fish Department opposes this plan because the agency manages by herd unit objectives which include consideration for sustained yield and because they are concerned by the increasing proportion of elk on the National Elk Refuge which come from Grand Teton National Park.

As we will see in Chapters 3, 7 and 8, the Park cull appears necessary to restore the historical distribution of elk in the herd unit, and to ensure a kill large enough to maintain the population below limits set by the Wyoming Game and Fish Department for winter feedground numbers.

Feeding programs for elk

Many of the difficulties managing elk in the Jackson elk herd ultimately stem from the winter feeding programs. If it were not for winter feeding, the Park hunt would not be necessary. The feeding program is expensive, costing over $\frac{1}{4}$ million annually. Nevertheless, because winter range has been usurped by cattle ranching and the town of Jackson, winter feeding is necessary to sustain the large elk herd.

Controversy surrounds the manner in which the elk are fed, both at the National Elk Refuge and the State feedgrounds in the upper Gros Ventre River valley. Local interest groups feel that it is necessary that adequate feed be provided so that the elk do not lose weight during winter. Yet, maintenance of weight is not necessary to ensure survival or calf production. Local interest groups also express concern that pelleted alfalfa fed since 1975 on the Refuge is inferior to baled hay because it does not offer adequate roughage. However, extensive feeding trials have shown that elk actually maintain a higher level of nutrition on pelleted hay than on baled hay. In some years there is public clamor over the date for initiation of feeding. For example, in 1986 ranchers on the upper Gros Ventre complained that supplemental feeding was initiated too late because elk were feeding on private haystacks.

As will be developed more thoroughly in Chapter 5, Refuge and State feeding programs appear more than adequate. Overwinter losses on the National Elk Refuge average only 1.2% of the herd. As such, overwinter survival in the Jackson elk herd is considerably higher than for herds using natural winter ranges. A winter feeding program which provided a predetermined quantity of feed each winter irrespective of elk numbers would more closely resemble natural conditions. However, this would result in greater mortality on the Refuge during severe winters, and strong public empathy for winter-fed elk currently precludes this as a management option.

Herd size and harvest

Low numbers on the National Elk Refuge in 1984 created an enormous outcry from local residents, hunters, guides and outfitters. Attempts were made to blame the Refuge feeding program. Also, it was claimed that additional feedgrounds were necessary, more hay should be fed, baled hay was superior to the new pelleted alfalfa hay, etc.

The Wyoming Game and Fish Department was first to admit that culling levels had been excessive during the previous decade and that the herd had simply been reduced by overharvest. As a consequence of public pressure, hunting seasons were shortened and fewer hunting permits were issued. The herd responded quickly such that by 1987 the number of elk on the Refuge had already exceeded the quota of 7500 elk, and by 1989 numbered 9500 elk.

Range managers and agricultural interests claim that the elk herd is too large and to protect range resources and aspen in Jackson Hole, the elk herd should be reduced substantially (Beetle, 1979; Kay, 1985). Other factions desire increased elk numbers to enhance tourism and hunting opportunities.

There are no data which support claims that elk numbers are too high for the range resources available, and I find no evidence that the herd should be reduced to protect seasonal ranges. Nevertheless, there is solid evidence that elk suppress aspen regeneration in the vicinity of winter feedgrounds. Aspen are desirable trees that add considerable color to the autumn landscape. Areas in the vicinity of winter feedgrounds, e.g., in the upper Gros Ventre valley, will become sacrifice areas devoid of aspen unless extreme measures are taken to protect aspen stands. Burning enhances aspen sprouting and shoot growth; however, burned areas must be large or elk will converge on burned areas to forage and consequently prevent aspen regeneration (DeByle and Winokur, 1985).

In this book I will detail the various issues relating to management of the Jackson elk herd. I will then evaluate technical data relating to the migration, population biology, feeding program, and habitat ecology of elk to offer a basis for evaluating these issues. Although the data base for the Jackson elk herd is among the most extensive for any wild ungulate population, management decisions regarding wildlife are often based on socio-political considerations rather than biology. All that a wildlife biologist can do is to provide the best information possible, and a professionally-based opinion of the best management strategy. Despite frustrations created by the socio-political arena, an informed public and government will certainly make better management decisions.

Summary

1. The aim of this book is to synthesize information on the ecology and management of the Jackson elk herd in northwestern Wyoming. The herd is intensively managed, with elk being fed in winter and extensively culled by hunters in autumn.
2. The herd averages approximately 11 000 to 14 000 animals, ranging over an area of nearly 6000 km². The
3. The Jackson elk herd is migratory with principal summer ranges in Yellowstone National Park, Grand Teton National Park, the Teton Wilderness and other areas of Bridger–Teton National Forest.
4. Management of the herd is complicated because several federal and state agencies are involved, each with different management philosophies and priorities.
5. Winter feeding of the Jackson elk herd is justified because elk have been displaced from their limited winter range by cattle ranching and the town of Jackson.
6. A controversial hunt for elk occurs annually within Grand Teton National Park, which is necessary because of the winter-feeding program.
7. Escalating mineral, oil and gas development is the most serious threat to the viability of the Jackson elk herd.

Chapter 2.

Migration and seasonal distribution

Seasonal migrations of elk between high elevation summer ranges in southern Yellowstone National Park and winter range on the National Elk Refuge are among the most extensive for any elk population (Murie, 1951c; Anderson, 1958; Craighead *et al.*, 1972; Adams, 1982), and are among the most spectacular for any terrestrial large mammal. Contenders include migrations of barren-ground caribou (*Rangifer tarandus*) in the Arctic (Hemming, 1971), and wildebeest (*Connochaetes taurinus*) in East Africa (Sinclair, 1985). In each instance, the reason for these migrations is seasonal variation in climate which creates spatial variation in food abundance or availability (Baker, 1978).

For elk of the Jackson herd unit, winter snow accumulation on summer ranges appears to be the principal factor causing migration with much heavier snow at higher elevations and a gradient of increasing winter snow accumulation from south to north in Jackson Hole (Figure 2.1). Average maximum depth of snow on the ground in January is only 42.9 ± 3.77(SE) cm at Jackson near the National Elk Refuge but averages 143 ± 6.95(SE) cm at the Snake River weather station near the South Gate of Yellowstone National Park.

Detailed understanding of distribution and migration are important to management. This is particularly true for autumn migration of the Jackson elk herd when much of the annual hunter kill of elk occurs. Strategic timing and location of the harvest along migration routes is required to ensure the desired distribution of harvest amongst the various herd segments, particularly those summering in Yellowstone National Park, which are inaccessible to hunters until the migration begins.

In this chapter I (1) evaluate techniques which have been employed to monitor the distribution and migration of the Jackson elk herd, (2) test the hypothesis that snowfall influences the timing and spatial pattern of fall migration, (3) document changes in distribution and migration which have

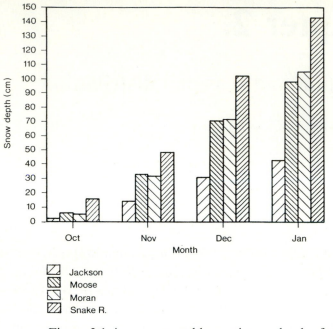

Figure 2.1 Average monthly maximum depth of snow on the ground (cm) for October, November, December, and January (1948–85) at four weather stations in Jackson Hole: (1) Jackson, (2) Moose, (3) Moran, and (4) Snake River.

occurred since the expansion of Grand Teton National Park in 1950, and (4) evaluate factors contributing to these changes. Finally, I summarize available information on interchanges between the Jackson elk herd and surrounding elk populations.

General distribution patterns

Boundaries for the Jackson elk herd unit were established by the Wyoming Game and Fish Department to encompass a distinct group of elk with hopefully no more than about 10% interchange with surrounding herds (Gasson, 1987: 10). To simplify complex geographic details, I identify three main groups within the Jackson elk herd based upon summer distribution (Figure 2.2). These are broad groupings and there is clearly movement of elk between them. Most elk from these three groups winter on the National Elk Refuge or on one of three state feedgrounds in the upper Gros Ventre River valley, and 10–20% of the herd winters on natural winter range on the Bridger–Teton National Forest. Although summer ranges are extensive and encompass approximately 4740 km², winter ranges are restricted to less than 455 km². Consequently, these winter ranges are generally more critical in the management of the herd.

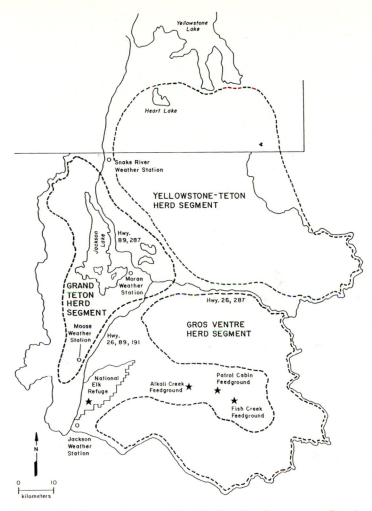

Figure 2.2 General geographic distribution in summer for the three main herd segments, and locations for feedgrounds and weather stations in the Jackson herd unit.

The Yellowstone–Teton Wilderness herd segment

This is the well-known migratory population which winters on the National Elk Refuge and the Gros Ventre feedgrounds, as well as on winter ranges along the Buffalo Fork valley, Spread Creek and along the Gros Ventre River. These elk summer on high-elevation ranges in southern Yellowstone National Park and the adjacent Teton Wilderness to the south. Migration usually occurs during May and June and during October and November, entailing a maximum straight-line distance of over 100 km (Figure 2.3). During July and August, the principal summer range is along several north–south ridges: Big Game Ridge, Chicken Ridge, Red Creek Ridge, Huckleberry Ridge,

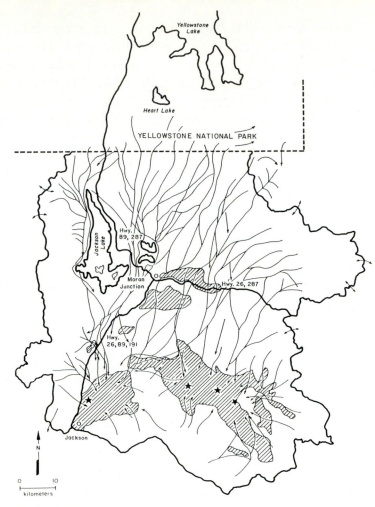

Figure 2.3 Autumn migration corridors determined by aerial survey. Hatched areas are elk winter range and stars indicate locations of winter feedgrounds.

and Two Ocean Plateau (Anderson, 1958; Cole, 1969). Elk from the Jackson herd seldom migrate farther north than a line running west from the south end of the South Arm of Yellowstone Lake just north of Heart Lake and continuing west to northern Pitchstone Plateau (Craighead *et al.*, 1972).

The Grand Teton National Park herd segment

Prior to expansion of Grand Teton National Park in 1950, few elk summered there, probably because they had been displaced by agricultural development in the valley (Anderson, 1958; Cole, 1969). Since then, however, this herd segment has increased and now probably constitutes one third to one half of the elk wintering on the National Elk Refuge. Although sometimes referred to as the 'non-migratory' segment of the herd, some elk summering in the north end of Grand Teton National Park, e.g., in the Berry Creek area, migrate over 70 km to and from the National Elk Refuge. Most elk that summer in the Park are in the Snake River valley between Signal Mountain and Timbered Island, approximately 30 km north of the National Elk Refuge.

The Gros Ventre Forest herd segment

This segment is the least well defined of the three. It encompasses elk that summer in the vast area east of Grand Teton National Park and south of Highway 26/287, much of which is drained by the Gros Ventre River, Spread Creek, and their tributaries (Figure 2.2). The elk are scattered over a large area and typically do not reach the high densities which can occur on the other two summer ranges. Many elk that summer here winter at feedgrounds in the upper Gros Ventre River valley, but some may winter at the National Elk Refuge. Interchanges between feedgrounds are common (Anderson, 1958).

Winter distribution

During winters when supplemental feed is provided, most of the elk herd concentrates on the three state feedgrounds and the National Elk Refuge. In mild winters when supplemental feeding is not necessary, elk are distributed more widely over the crucial elk winter range within the Jackson herd unit (Figure 2.3). These winter ranges may reflect prehistoric winter ranges, except that wintering elk probably moved further south along the Snake River downstream from the present location of the town of Jackson (Murie, 1951c: 60; Anderson, 1958: 53–60; Cole, 1969). Earlier observers stated that prior to 1913 many elk from the Jackson herd migrated into the Green River drainage (Murie, 1951c: 60; Anderson, 1958), but evidence for this migration pattern is poor and largely unsubstantiated (Cole, 1969).

Techniques for studying migration and distribution

As Peek (1985) emphasizes, there are no reliable techniques for censusing elk in forested habitats; therefore indirect methods must be used to estimate patterns of distribution, migration and abundance. Several techniques have been employed to study the distribution and migration of the Jackson elk herd. These include aerial surveys (Anderson, 1958), neck banding (Knight, 1966), ear tagging (Straley, 1968), radio telemetry (Long *et al.*, 1980), trail traffic counters (Wood & Roby, 1975), and track counts (Anderson, 1958; Cole, 1969).

Aerial surveys have been used to count elk in Jackson Hole since 1932

(Anderson, 1958), but methodology has varied from year to year so that data are often not directly comparable among surveys. During the 1950s, six summer trend count transects were established (Wilson, 1958: 6), but the number of elk counted was extremely variable and these transects were later discontinued.

In recent years there has been considerable research on aerial survey methods for estimating large mammal populations (Caughley, 1974; Cook & Jacobson, 1979; Samuel & Pollock, 1981), and methods have been developed for animals such as elk where a fraction of the population is not seen (Samuel, Garton, Schlegel & Carson, 1987). Standard line transect methods are not appropriate for aerial survey of elk populations because all animals 'on the transect line' must be counted (Burnham, Anderson & Laake, 1980). However, the fraction of elk missed during an aerial survey may be estimated by observing marked elk from the air from a known number of marked animals (Samuel *et al.*, 1987). Unfortunately, the fraction of elk missed during an aerial survey is usually unknown, and in forested habitats this fraction may be very large and highly variable (Buechner, Buss & Bryan, 1951; Lovaas, Egan & Knight, 1966).

Even though aerial surveys are often impractical for estimating elk populations, flights can be very useful for locating migration routes when tracks can be seen in the snow (Robel, 1960). This method was used to compile Figure 2.3. In addition, the Wyoming Game and Fish Department uses aerial surveys to locate elk wintering off feedgrounds.

During 1959, 588 elk were marked with white collars on the National Elk Refuge to estimate the number of elk using various summer ranges. Subsequently, many of these animals were observed in southern Yellowstone National Park and a Lincoln-Petersen estimator was employed to estimate that 4100 elk from the Jackson elk herd summered in Yellowstone National Park in 1964 (Cole, 1965, 1969). Of these, 63 of 67 observations were on Big Game and Chicken Ridges and only four were further east. Since these first studies, various numbers of elk have been marked with visible collars to document distribution on summer ranges (Cole, 1969; Martinka, 1969).

From 1943 to 1989, over 11 000 elk have been ear-tagged on the National Elk Refuge and Gros Ventre feedgrounds. When hunters kill an ear-tagged elk they are expected to report ear-tag numbers to the Wyoming Game and Fish Department. These data have provided detailed information on fall migration and distribution, and interchanges with other herds (Straley, 1968).

Two studies employed radio telemetry to monitor movements of elk in the Jackson elk herd: (1) a study of elk/logging relationships in the Mt Leidy Highlands where 21 elk were radio-collared (Long *et al.*, 1980), and (2) a study of 97 elk radio-collared on the National Elk Refuge (B. Smith, unpublished).

Although radio telemetry allows collection of extensive information on individual animals and detailed description of movements, radio transmitters are expensive and usually only a few animals can be monitored at one time. Consequently, sample sizes are typically too small to provide much confidence in inferences about herd distribution.

A trail traffic counter was used in 1965 and 1975 to count elk use of two heavily-used elk trails on the west shore of Jackson Lake (Cole & Yorgason, 1965; Wood & Roby, 1975). Although counters may be useful for monitoring selected heavily-used trails, on one of the trails monitored in 1975, elk were able to walk around the counter and thereby some were not counted.

Finally, starting in 1945, the Wyoming Game and Fish Department and the National Park Service have collaborated to count elk tracks crossing snow-covered road transect routes during fall migration (Anderson, 1958; Cole, 1969). Track transects were established along roads from Beaver Creek in Grand Teton National Park to Togwotee Pass on the eastern boundary of the herd unit (Cole, 1969). Initially, data were recorded separately for 14 road segments, but after the new highway was constructed in 1959, some of these were combined into the present eight transects (Figure 2.4).

Transects are driven beginning as early as 1 October if sufficient snow is present, and the number of elk tracks crossing the highway from north to south is recorded. Tracks heading south to north are subtracted from the count. After counting, if snow plows are not likely to obliterate the tracks, this is accomplished with a garden rake or a drag towed behind a vehicle to ensure that tracks are not recounted during a subsequent survey (Anderson, 1958; Yorgason & Cole, 1963). When several elk cross the road at the same place, the trail is backtracked to where the tracks separate and can be counted. Sometimes even backtracking may not result in an accurate count, particularly when many elk follow the same trail (Figure 2.5). Surveys continue through November, and in some years into December or even January if migration is still occurring.

These track counts have documented shifts in the spatial distribution and timing of migration from year to year. However, conditions for conducting track counts varied; for example, in 1961 track counts were almost totally unsuccessful because the migration occurred in October before there was enough snow to monitor track crossings (Bendt & Yorgason, 1961). For various reasons, the number of days that transect counts were conducted each year between 1950 and 1984 varies (Figure 2.6), except 1955, 1956, and 1961 for which there are no data. Consequently, the track counts show enormous variance among years, presenting some difficulties for data analysis. Furthermore, some transects were covered more times than others, often because snow cover did not exist at lower elevations, but was adequate at higher elevations.

Figure 2.4 Track transect routes for monitoring autumn migration. From west to east, the eight main transects are (1) Burned Ridge, (2) Snake River, (3) Pacific Creek, (4) Buffalo River, (5) Boundary, (6) Blackrock, (7) Four-mile Meadow, and (8) Togwotee. Additional transects monitored since 1976 along Highway 26 north of Jackson are also mapped, and labelled to coincide with Table 2.6. Feedground locations are starred.

Accepting that there will be substantial sampling variance, the data base is nevertheless very extensive and appears to offer useful information once subjected to appropriate statistical analysis. As Wilson (1958) suggested, the track counts provide the best trend index presently available for monitoring migration of the northern segment of the Jackson elk herd.

Beginning in 1976 additional track transects (see Figure 2.4) were initiated to monitor the movement of elk from Grand Teton National Park onto the National Elk Refuge. When combined with data from the original eight transects, these new transects permit monitoring the movement of elk onto the National Elk Refuge.

Interchanges with adjacent elk herds

Ear-tag returns (Table 2.1) and neck-banding studies of elk from Jackson Hole and surrounding areas have documented interchange between herds (Anderson, 1958; Cole, 1969; Craighead *et al.*, 1973). Generally, only a

Figure 2.5 Single-file migration of elk can make track counts difficult. Photo courtesy of the National Elk Refuge.

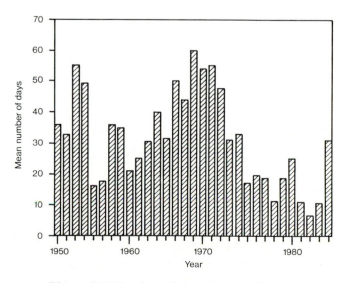

Figure 2.6 Number of days that track transects were counted, 1950–84, excluding 1955, 1956 and 1961 for which data do not exist.

Table 2.1. *Ear-tag returns from Jackson elk herd, 1951–85*

Year	Number of recoveries elk tagged in Jackson herd unit	Number of recoveries outside Jackson herd unit	Percent elk dispersing outside Jackson herd unit	Recoveries elk tagged outside Jackson herd unit
1951	126	5	4.0%	0
1952	22	0	0.0%	0
1953	5	0	0.0%	0
1954	0	0	0.0%	0
1955	1	0	0.0%	0
1956	0	0	0.0%	0
1957	6	1	16.7%	0
1958	4	0	0.0%	18
1959*	32	1	3.1%	0
1960	95	4	4.2%	0
1961	106	13	12.3%	2
1962	86	3	3.5%	0
1963	183	12	6.6%	1
1964	100	0	0.0%	1
1965	157	3	1.9%	3
1966	142	3	2.1%	2
1967	16	0	0.0%	0
1968	56	3	5.4%	2
1969	106	2	1.9%	3
1970	185	5	2.7%	2
1971	107	6	5.6%	8
1972	92	2	2.2%	4
1973	155	4	2.6%	5
1974	78	6	7.7%	3
1975	123	11	8.9%	2
1976	34	3	8.8%	0
1977	87	5	5.7%	2
1978	76	5	6.6%	0
1979	83	2	2.4%	0
1980	36	2	5.6%	1
1981	37	2	5.4%	3
1982	17	1	5.9%	0
1983	48	0	0.0%	1
1984	11	1	9.1%	1
1985	14	2	14.3%	0
Totals	2426	107	4.4%	64

* Beginning in 1959, recoveries for elk killed each year. Prior to 1959, recoveries are for elk tagged each year.

small proportion of marked elk have emigrated to surrounding herd units. Of 2426 ear-tag recoveries from elk tagged within the Jackson herd unit, 107 were recovered outside the Jackson herd unit. This constitutes an emigration rate of only 4.4%.

Interchange between Shoshone River, northern Yellowstone, and Jackson herds is remarkably small given that elk from all three populations use common summer ranges in southern Yellowstone National Park (Anderson, 1958; Yorgason, 1966; Craighead *et al.*, 1973). Movements by some individuals have been remarkable, although most elk recovered outside the herd unit are in adjacent areas east or south. Recoveries have come from numerous places 200–400 km away including Idaho; Montana; northern Yellowstone National Park; Dubois, Wyoming; Greys and Green Rivers in Wyoming; and in the vicinity of South Park feedground 10 km south of Jackson.

During recent studies of the Sand Creek elk herd immediately west of the Jackson herd in Idaho, two out of 53 radio-collared elk moved into Jackson Hole during summer (Brown, 1985). One bull wandered as far east as the south entrance to Yellowstone National Park within the Jackson herd unit. Interchange between winter ranges for these animals is further reinforced by movements of a cow elk which was radio-collared at the Sand Creek winter range and subsequently was observed 100 km northeast near West Yellowstone, Montana, and finally moved another 125 km further south to spend summers near Teton Pass and winter on the South Park feedground south of Jackson.

A substantial fraction of the elk that winter at Sand Creek in Idaho summer on the Pitchstone Plateau in southwestern Yellowstone National Park. This area is adjacent to the Jackson herd unit, and Brown (pers. comm.) speculated that a greater proportion of elk from the Pitchstone Plateau now migrate to the National Elk Refuge than when his study began in 1980. There is no documentation for such a pattern; in fact, since 1976 it seems probable that a greater fraction of the elk summering on the Pitchstone Plateau may now winter in Idaho than on the National Elk Refuge. Cole (1969) estimated that 800 elk from the Jackson elk herd summered on the Pitchstone Plateau in 1964, which was reinforced by track counts along the Grassy Lake road in the Rockefeller Parkway (Yorgason & Cole, 1967). Since 1976, using aerial surveys and track counts, G. Roby (pers. comm.) has been unable to document a sizable migration of elk from the Pitchstone Plateau toward Jackson. Brown (pers. comm.) concludes that there is inadequate evidence to confirm that an increasing proportion of elk from the National Elk Refuge now migrate into Idaho, but it is certainly possible.

Timing of migration

Elk usually leave winter feedgrounds during April or early May (Figure 2.7; Table 2.2). Elk migrate north, with pregnant females interrupting migration briefly to give birth in traditional calving areas (see Chapter 4). Males and nonbreeding females continue migration on toward the summer ranges. Females with calves follow 2–3 weeks later (Cole, 1969; Anderson, 1958; Murie, 1951c; Altmann, 1952). Elk summering on high-elevation ranges, for example, in southern Yellowstone National Park and the Teton Wilderness, may follow the receding snow line, typically following plant green-up (Anderson, 1958; Cole, 1969; Sweeney, 1976).

Figure 2.7 Exodus of elk from the National Elk Refuge, 27 April 1985. Photo courtesy of B. Smith, National Elk Refuge.

Summer ranges are used until at least mid-August, followed by the rut in mid-September. The southward autumn migration typically peaks during October or November, as reflected by their crossing of road transects en route to winter ranges or feedgrounds. Arrival of elk onto the National Elk Refuge in fall is monitored by Refuge personnel (Tables 2.2 and 2.3).

If track counts accurately describe the timing of migration, peak track counts should reasonably conform with the arrival of elk on the National Elk Refuge.

One complication in evaluating such a pattern is that track transects are covered with varying frequency from year to year, and within years. The number of tracks counted is positively correlated with the mean number of days that transects were counted from year to year ($r = 0.547$, $N = 32$, $P < 0.001$). Likewise, grouping records of track counts into semimonthly periods to correspond with the periods for counts on the National Elk Refuge, the number of tracks counted per semimonthly period is positively correlated with the mean number of days that transect counts were conducted ($r = 0.508$, $n = 153$,

Table 2.2. *Bimonthly maximum counts of elk on the south two-thirds of the National Elk Refuge, 1958–85*

Year	Sept. 15–30	Oct. 1–15	Oct. 16–31	Nov. 1–15	Nov. 16–30	Dec. 1–15	Dec. 16–31	Jan. 1–15 (yr+1)	Date of spring exodus
1958		300	809	1500	5000	4200	4500	*	
1959		775	1550	2533	4000	3800	4200	*	
1960		1100	2500	3500	5800	5850	6200	*	
1961		1726	3470	4500	*	6200	*	*	
1962		2450	2500	2490	2600	*	4036	*	
1963		145	728	5000	7000	6225	*	*	
1964		145	1000	1800	2900	4500	6000	*	
1965		0	375	521	2300	3900	*	*	
1966		36	300	1540	4000	*	*	*	
1967		0	200	2025	3900	*	*	*	
1968		0	0	2600	6000	6700	7000	7195	
1969		0	0	300	975	3000	3900	7360	
1970		0	125	2100	4100	7515	7630	8495	
1971		0	70	2000	4075	6600	6610	6590	
1972	22	18	40	250	1700	4700	5453	6035	11 May 1973
1973	12	12	12	5000	7200	7500	7400	7100	4 May 1974
1974	47	76	45	100	2000	5050	5100	6201	
1975	7	44	1600	3115	7000	7106	7500	7950	
1976	20	20	77	5	35	50	900	4295	<19 April 77
1977	0	0	93	250	6010	8250	8210	8600	
1978	137	130	130	0	3945	7000	7500	7150	13 May 1979
1979	147				600	*	6805	6299	5 May 1980
1980	57				36	4470	5000	6180	<10 April 81
1981		13	7	2	3475	5010	5500	5900	27 April 1982
1982	86			415	3350	5620	5662	5397	10 May 1983
1983				1775	4630	4450	4578	4750	
1984			7	3425	4690	5659	4775	5105	5 May 1985
1985	14	4	37	2170	5668	5562	5445	6480	13–19 April 86

* no data

$P< <0.001$). Therefore, I weighted track counts by dividing by the number of days that each transect count was conducted during each semimonthly period.

For each year, I compared the semimonthly period during which peak weighted track counts occurred with the semimonthly period during which the maximum increase in Refuge counts occurred. I then scored the time lag in the coincidence between these peak counts for (1) weighted track counts recorded on transects within Grand Teton National Park (Burned Ridge through Boundary transects), and (2) weighted track counts recorded on transects east of the Park on the National Forest (Blackrock, Four-mile Meadow and Togwotee transects).

In most years the peak in weighted track counts for both Park and Forest track transects coincided with the same semimonthly interval during which the maximum influx of elk occurred on the National Elk Refuge (Figure 2.8).

Table 2.3. *Fall migration reflected by date when 500 or more elk were recorded on the National Elk Refuge.*

Year	Date	Year	Date	Year	Date
1915	Dec 10	1940	Oct 29	1965	Nov 05
1916	Dec 21	1941	Nov 06	1966	Nov 08
1917	No data	1942	Nov 04	1967	Nov 06
1918	No data	1943	Oct 22	1968	Nov 08
1919	Nov 03	1944	Nov 06	1969	Nov 18
1920	Nov 22	1945	Nov 08	1970	Oct 29
1921	Nov 24	1946	Oct 21	1971	Nov 05
1922	No data	1947	Nov 08	1972	Nov 26
1923	Dec 08	1948	Oct 10	1973	Nov 05
1924	Nov 06	1949	Oct 15	1974	Nov 20
1925	Oct 10	1950	Oct 08	1975	Oct 28
1926	Oct 28	1951	Oct 20	1976	Dec 22
1927	Nov 10	1952	Oct 13	1977	Nov 21
1928	Nov 22	1953	Oct 15	1978	Nov 24
1929	Oct 30	1954	Oct 05	1979	Nov 24
1930	Oct 27	1955	Oct 02	1980	Dec 03
1931	Oct 22	1956	Oct 25	1981	Nov 25
1932	Oct 15	1957	Oct 07	1982	Nov 22
1933	Oct 18	1958	Oct 21	1983	Nov 15
1934	No data	1959	Oct 22	1984	Nov 14
1935	Nov 15	1960	Oct 05		
1936	Dec 08	1961	Sep 13		
1937	Nov 23	1962	Oct 12		
1938	Nov 05	1963	Oct 22		
1939	Jan 03	1964	Oct 25		

However, peak weighted counts from Park transects are more likely to precede the peak influx on the National Elk Refuge, whereas peak weighted counts on Forest transects are more likely to occur at the same time or even after the peak arrivals on the Refuge. The null hypothesis that Park and Forest transects have the same frequency of years during which the timing of migration was before, coincident with, and after the peak migration onto the National Elk Refuge was rejected ($\chi^2 = 6.46$, $P < 0.05$).

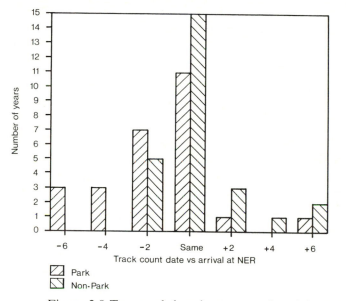

Figure 2.8 Two-week lags between peak weighted track counts and peak arrivals on the National Elk Refuge (1958–84). Park transects include Burned Ridge through Boundary, whereas non-Park transects are Blackrock through Togwotee.

Many of the elk summering in the central valley portion of Grand Teton National Park are south of the track transect route, and there is no assurance that elk crossing Park transects are elk which have summered within the Park. Yet, field observations on summer range generally note earliest migration occurring among upper Berry Creek areas in northwestern Grand Teton National Park (Bendt & Yorgason, 1961: 7). Another possible explanation for the difference in timing of migration across Park and Forest transects is that migration may be faster within the Park because there is less security cover than in the forested areas further east. Radio-telemetry studies (B. Smith, unpublished) show that the duration of migration is shorter for elk summering in Grand Teton National Park (probably because they are closer to the National Elk Refuge), and they cross transects and arrive at the Refuge sooner than other elk.

Based upon observations of migrating elk, it is thought that the earliest elk to migrate onto the National Elk Refuge are elk from Grand Teton National Park (Smith, 1985). It is this pattern that justifies an early hunt on the National Elk Refuge; that is, the timing of the hunt is set in an attempt to kill mostly Park elk, particularly those that make the longest use of the Refuge. During the early 1960s the early hunt on the Refuge was killing a portion of the Park herd that typically moved onto the Refuge in mid-October (F. Petera, Wyoming Game and Fish Department, pers. comm.). As can be seen in Table 2.3, since 1965, fewer than 500 elk have been on the Refuge by 1 November each year with the exception of 1975 when 500 appeared on 28 October. This may suggest that early-season hunts on the National Elk Refuge eliminated herd segments entering the Refuge early in the fall.

Influence of climate on migration timing

Others have hypothesized that timing of migration for the Jackson elk herd in fall is stimulated by snowfall, and migration during spring is delayed by late snow melt (Anderson, 1958; Wilson, 1958; Cole, 1969). Evidence for this has been largely anecdotal. In November 1966, 2500 elk moved onto the National Elk Refuge after a heavy snowfall (National Elk Refuge, 1966). In 1944, relatively late snow accumulations were believed to have delayed the migration of elk, which remained at high elevations longer than usual (Murie, 1951b). Migration of elk in other areas is influenced by snow depth (Dalke *et al.*, 1965a; Sweeney, 1976). Migration of elk from southeastern Yellowstone National Park into the Shoshone River drainage occurred in response to approximately 20 cm of snow (Rudd, Ward & Irwin, 1983). Here I analyze the effect of climate on migration of the Jackson elk herd.

Climatic data

Several weather stations exist in Jackson Hole, but only four provide long-term and consistent records of temperature, precipitation and snow depth. These are, from south to north, (1) Jackson at 1904 m, (2) Moose at 2020 m, (3) Moran station located 5 km northwest of the village of Moran at Jackson Lake Dam (2054 m), and (4) Snake River station located near the south gate of Yellowstone National Park at 2098 m (Figure 2.2).

Studies of climatic variability in Grand Teton National Park show that the four stations adequately represent the range of climatic variation at low elevations in the valley (Dirks & Martner, 1982). Typically, temperatures are lower and precipitation and snow depth are greater along a south-to-north gradient through Jackson Hole.

Data from three of the weather stations exist for as long as statistics on the elk herd have been gathered. However, recording of climatic data did not begin at Moose until October 1935. But even at stations which have extensive runs of data, data are missing for some months throughout the period of record. Nevertheless, climatic variables at all stations are highly intercorrelated. Thus missing values can be extrapolated with reasonable accuracy from records from the other stations.

Dummy variable regression

The timing and intensity of track counts varies considerably among years. Some counts were initiated on 1 October and continued through January whereas others were only conducted during November. The fact that counts were not conducted during a particular period does not imply that no migration occurred then; only that there is no record of whether or not tracks crossed the transects. Sometimes counting was impossible because there was no snow on a particular transect during a two-week period. This structure lends itself to dummy variable regression where each semimonthly period (t_i) is assigned a value of 0 or 1 and is entered into model estimation only when there are data for the semimonthly period.

The model is

$$y_i = t_1 + t_2 + \ldots + t_7 + S_{\text{Oct}} + S_{\text{Oct-Nov}}$$
$$+ S_{\text{Oct-Dec}} + S_{\text{Nov}} + S_{\text{Nov-Dec}} + S_{\text{Dec}} + \epsilon \qquad (2.1)$$

where y_i is the proportion of total yearly track counts occurring during a semimonthly period (t_i), where t_1 is 1–15 October, t_2 is 16–31 October, ... and t_7 is 1–31 January of the following calendar year. The S_i's are the respective monthly maximum depth of snow on the ground in cm at the Moran, Wyoming weather station. When two months are subscripted under a snow variable, the first month refers to the month for snow depth and the second variable refers to the month of the track counts being affected by the snowfall. For example, $S_{\text{Oct-Nov}}$ refers to October maximum snow depth as it influences the weighted proportion of tracks counted during the two November track count periods. For all time periods other than November, $S_{\text{Oct-Nov}} = 0$. Finally, ϵ is an error term.

The model is fitted using weighted regression where weights are inversely proportional to variance. Since the y_i's are proportions, the appropriate weights are

$$w_i = \frac{n}{[y_i(1 - y_i)]} \qquad (2.2)$$

for n equal to the mean number of days that the transects were counted during the semimonthly period.

Selected models showing the role of snow depth on migration timing are presented in Table 2.4 for (1) pooled track transects within Grand Teton National Park, (2) pooled track transects for National Forest transects east of the Park, and (3) all transects combined. These results show that on Park transects, snow depth in October is inconsequential to the timing of migration,

Table 2.4. *Dummy variable regression (see equation 2.1) of the timing of elk migration across track transects in (a) Grand Teton National Park, (b) east of the Park, and (c) pooled Park and Forest transects. Dependent variable, y, is the weighted proportion of track counts occurring during each semimonthly period,* t_i.

(a) Park transects. $r=0.671$, $n=155$, $\bar{y}=0.134$.

Variable	Coefficient	Std. error	t	P
$t_2 = 15–31$ Oct	0.095	0.035	2.75	0.007
$t_3 = 1–15$ Nov	0.146	0.061	2.41	0.017
$t_4 = 16–30$ Nov	0.158	0.065	2.44	0.016
$t_5 = 1–15$ Dec	0.278	0.062	4.50	0.000
$t_6 = 16–31$ Dec	0.134	0.059	2.28	0.024
$t_7 = 1–31$ Jan	0.233	0.114	2.04	0.043
S_{Nov}	0.009	0.004	2.20	0.029
$S_{Nov–Dec}$	−0.010	0.004	−2.61	0.010

Analysis of variance

Source	Sum-of-squares	DF	Mean-square	F-ratio	
Regression	3.013	8	0.377	15.039	$P<0.001$
Residual	3.682	147	0.025		

(b) Park transects east of the Park. $r=0.71$, $n=131$, $\bar{y}=0.164$.

Variable	Coefficient	Std. error	t	P
$t_4 = 16–30$ Nov	0.238	0.060	3.95	0.000
$t_5 = 1–15$ Dec	0.366	0.079	4.64	0.000
$t_6 = 16–31$ Dec	0.062	0.073	0.86	0.393
$t_7 = 1–31$ Jan	0.491	0.186	2.65	0.009
$S_{Oct–Dec}$	−0.032	0.014	−2.25	0.026
S_{Nov}	0.008	0.003	2.72	0.007
$S_{Nov–Dec}$	0.005	0.005	0.96	0.340

Analysis of variance

Source	Sum-of-squares	DF	Mean-square	F-ratio	
Regression	4.455	7	0.636	17.985	$P<0.001$
Residual	4.388	124	0.035		

(c) *Pooled Park and Forest track transects.* r = 0.698, n = 155, ȳ = 0.127.

Variable	Coefficient	Std. error	t	P
t_2 = 15–31 Oct	0.072	0.029	2.50	0.013
t_3 = 1–15 Nov	0.150	0.051	2.94	0.004
t_4 = 15–30 Nov	0.201	0.056	3.58	0.000
t_5 = 1–15 Dec	0.306	0.056	5.42	0.000
t_6 = 16–31 Dec	0.133	0.051	2.59	0.011
t_7 = 1–31 Jan	0.325	0.119	2.73	0.007
S_{Nov}	0.004	0.003	1.16	0.247
$S_{Nov-Dec}$	− 0.009	0.003	− 2.45	0.016

Analysis of variance

Source	Sum-of-squares	DF	Mean-square	F-ratio
Regression	2.536	8	0.317	17.485 $P < 0.001$
Residual	2.665	147	0.018	

but heavy snow accumulation in November results in a high proportion of elk migrating during November. Heavy snow accumulation in November is concomitantly inversely correlated with the relative proportion of tracks counted in December or January. In other words, heavy snow accumulation in November results in peak migration in November and consequently fewer elk migrating later.

On Forest transects, maximum snow depth in October contributes to a reduction in late-season migration, resulting in few tracks during December and January. Migration during November is positively correlated with the accumulation of snow in November. In contrast, October snow depth contributes to relatively early migration (or at least reduced late migration) across transects in the Park. This occurs because the Forest transects include the Togwotee Pass transect, which is at higher elevation; migration late in the season across Togwotee is uncommon if snow accumulation is high.

Combined data for Park and Forest transects reflect the same patterns as for Park transects. In all cases, counts during 1–15 October (t_1) do not contribute significantly to the models. This may be partly due to the paucity of data from the first half of October when in many years snow is inadequate to conduct the transect counts. December snow depth does not contribute significantly to any of the models because in most years, peak migration occurred before December, and maximum December snow depth is highly correlated with November maximum snow depth ($r = 0.758$, $N = 25$, $P < 0.001$).

These results confirm that snow accumulation plays a major role in the timing of migration for elk in the Jackson herd unit. Because snow accumulates early in high-elevation areas north of Togwotee Pass, migration is typically

earliest there. Migrations through Grand Teton National Park are next, followed by migrations across the heavily forested transects east of Grand Teton
National Park. Because snow accumulation is unpredictable, migration timing
is also unpredictable and varies considerably from year to year.

Changes in migration and distribution

Individual elk often show fidelity to seasonal ranges (Knight, 1970;
Craighead *et al.*, 1973; Irwin & Peak, 1983b; Edge, Marcum & Olson, 1985).
Changes in migration pattern for the Jackson elk herd are believed to be a
consequence of various herd segments having been virtually eliminated by
hunting (Bendt, 1962). Additionally, Sweeney (1976) found that climatic variation could cause shifts in the spatial pattern of migration as well as timing of
migration. Therefore, in this section I test the null hypothesis that the spatial
pattern of migration is not correlated with various combinations of year-to-
year variation in hunter kill and snow accumulation during autumn.

The percentage of tracks counted along transects within Grand Teton
National Park increased between 1949 and 1957, and increased still more
between 1957 and 1964 compared to other transects outside the Park (Cole,
1969). However, this comparison did not consider differences among transects
in the number of days that transects were counted. One reason that variation in
counting intensity occurs between Park and Forest transects is because
transects from Burned Ridge to Buffalo River are usually counted by National
Park Service personnel, and eastern transects to Togwotee Pass are usually
counted by Wyoming Game and Fish Department personnel. A further confounding factor is that the Burned Ridge transect is often run for a shorter time
than other transects because the road is usually closed before termination of
track counts on other transects. Consequently, the intensity of coverage varies
among transects. As in the previous analysis, to study spatial variation in
migration pattern, I inversely weighted track counts by the total number of
days that transects were counted each year (C_i/w_i), where C_i is the number of
tracks on the ith transect and w_i is the number of days that the ith transect was
counted during each year. Then I calculated the weighted proportion of tracks
coming from each transect,

$$C'_i = (C_i/w_i)/\sum(C_i/w_i) \tag{2.3}$$

Plotting these weighted proportions of tracks counted reveals marked trends
(Figure 2.9). The pattern which Cole (1969) suggested is strongly supported by
the substantial increase in the weighted proportion of tracks across Park
transects versus those east of the Park. This is particularly true for the Burned
Ridge (BR) transect which monitors migration of elk following routes west of
Jackson Lake. Park transects are pooled in Figure 2.10.

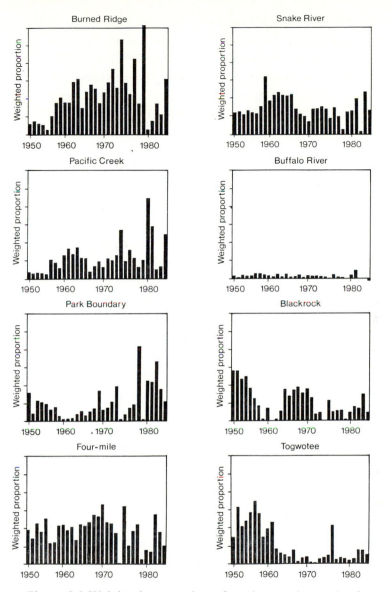

Figure 2.9 Weighted proportion of tracks crossing each of the eight transects, 1950–84. No count data exist for 1955, 1956 or 1961.

Perhaps most striking is the decline in use of the Togwotee transect, which was the most heavily used migration route prior to 1958. From 1958 to 1964 use diminished to a low level which has prevailed to the present (Figure 2.9). A peak appears in the relative proportion of tracks using the Togwotee transect in 1976. This is anomalous, because there was very little snow in 1976. Since the Togwotee transect is highest in elevation and therefore accumulated more snow, the few tracks that were recorded in that year occurred on this transect.

Patterns shown in Figure 2.9 suggested to me that the spatial distribution of migration might be more even among transects in more recent years. However, calculation of a Shannon-Weaver evenness index (Pielou, 1975) for each year shows that evenness actually declines in more recent years (Figure 2.11). The principal reason for this is smaller sample sizes in several recent years which generates high sampling variance in weighted proportions of track counts within transects.

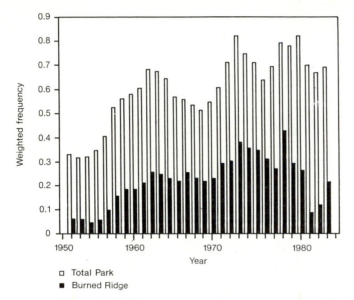

□ Total Park
■ Burned Ridge

Figure 2.10 Three-year moving average of weighted proportion of track counts summed over all transects within Grand Teton National Park, i.e., transects Burned Ridge through Boundary. Also plotted is the three-year moving average of weighted proportion of tracks crossing the Burned Ridge transect.

Trends illustrated in Figure 2.9 are of interest. Year-to-year variation in migration was similar for the Snake River, Pacific Creek, and Buffalo River transects, as it was for the Boundary, Blackrock, and Four-Mile Meadow transects. Statistical analysis of annual variation in migration pattern is presented in Appendix A. Hunter harvest along migration routes has major influences on migration across transects in subsequent years. Snow depth also plays an important role, whereby heavy snows result in a greater fraction of the migration crossing Boundary, Blackrock and Four-mile Meadow track transects, and a lower fraction crossing transects within Grand Teton National Park.

The effect of weather on the migration of elk is reinforced by the results of aerial surveys which monitored the autumn migration of elk from summer

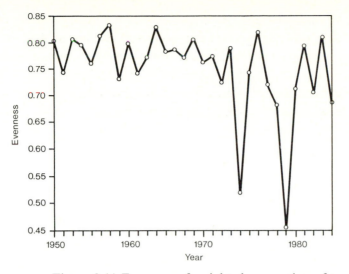

Figure 2.11 Evenness of weighted proportion of track counts across transects, 1950–84.

range near Heart Lake in Yellowstone National Park (Table 2.5). Depending upon weather conditions, elk were inclined to use Pilgrim Creek as a migration corridor or if migration was early, they were more likely to use Wolverine Creek and therefore emerge on a track transect further east.

I present one final data set on track counts. In Table 2.6 I summarize the observations of elk crossing road transects near the National Elk Refuge. Transects I and II monitor elk migrations coming from Grand Teton National Park rather directly onto the Refuge. Most of these elk summer in the vicinity of Stewart Draw and the J-Y Ranch within the Park (G. Roby, unpublished). Their migration route across the south end of hunt area 76B makes it unlikely that hunter harvest will occur until these animals are on the Refuge.

Interpretation of migration patterns

As postulated by others (Anderson, 1958; Bendt, 1962), both climate and hunter harvest are clearly important factors determining variation in the timing and spatial pattern of migration. Broad trends are largely a consequence of harvest patterns. The most pronounced trends occurred on the Togwotee and Burned Ridge transects.

Heavy hunter harvest in the late 1950s and early 1960s apparently depleted the herd segment which migrated across the Togwotee transect (Figure 2.12). Elk were hunted heavily in the Buffalo Fork headwaters, along the new Togwotee highway (completed November 1959), and along their migration route into Spread Creek and the Gros Ventre River drainage where road access increased markedly at this time. Also, in these early years, reductions of elk in Grand Teton National Park were usually later because the main objective for the Park hunt was to harvest elk migrating from summer ranges in Yellowstone National Park. The timing of migration for Togwotee transect elk made them susceptible to Park harvests as well. The net result was heavy harvest pressure on these elk throughout the entire season and a pronounced decline during 1958–65, followed by sustained low numbers since then. Extensive road construction in the area for logging has helped to sustain low elk populations (Johnson, 1976).

Track counts on the Burned Ridge transect increased consequent to enlarge-

Table 2.5. *Autumn migration route used by Heart Lake area elk, as monitored by aerial survey (G. Roby & R. P. Wood, unpublished). Elk use Pilgrim Creek as principal migration route during late migrations, but use Wolverine Creek if migration occurs early. Maximum depth of snow on the ground is recorded at Moran, Wyoming.*

Year	Aerial flight date	Migration route	Tracks on Snake River transect	Tracks on Pacific Creek transect	Maximum depth of snow, cm	
					October	November
1965	Nov 22	Pilgrim Creek	1913	946	1	36
1966	Nov 14	Pilgrim Creek	1139	578	3	30
1967	Oct 30	Wolverine Creek	577	170	5	25
1968	Nov 7	Wolverine Creek	869	574	1	36
1969	Nov 21	Pilgrim Creek	875	877	3	10
1970	Nov 3	Pilgrim Creek	315	305	5	43
1971	Nov 23	Pilgrim Creek	821	695	20	53
1972	Nov 7	Pilgrim Creek	623	459	3	13
1973	Nov 10	Pilgrim Creek	825	722	5	38
1974	Dec 2	Pilgrim Creek	300	578		25
1975	Oct 29	Pilgrim Creek	347	406	28	69
1976	Dec 29	Pilgrim Creek	325	361	0	3
1977	Nov 10	Pilgrim Creek	187	232	0	61
1978	Nov 16	Pilgrim Creek	48	134	0	43
1979	Nov 18	Wolverine Creek	416	235	0	18
1980	Nov 20	Pilgrim Creek	148	540	0	15
1981	Dec 6	Wolverine Creek	287	426		58
1982	Nov 15	Pilgrim Creek	28	120		28
1983	Nov 20	Pilgrim Creek	558	174	0	43
1984	Oct 29	Wolverine Creek	428	637	0	64

ment of the Park in 1950 due to the elimination of hunting in 1950 and cattle grazing west of the Snake River in 1958 (see Cole, 1969; Martinka, 1969). It is interesting to note the decline in use of the Burned Ridge transect in recent years, possibly due to the herd reduction program in Grand Teton National Park. However, the exceptionally low count on the Burned Ridge transect in 1980 can be attributed to very light snowfall in the valley; as a consequence, many elk had crossed the transect before snow was adequate for tracking.

Summary
1. Elk in the Jackson herd migrate up to 100 km from summer ranges in Grand Teton National Park and Bridger-Teton National Forest to winter ranges and feedgrounds in southern portions of the herd unit. Over 11 000 elk have been ear-tagged in the Jackson herd unit over the past 40 years. Returns of tags from hunters show that few elk move out of the herd unit.
3. Timing and routes of migration are substantially influenced by snow accumulation in fall and snow melt in spring.
4. Counts of elk tracks in the snow show that migration patterns have shifted substantially due largely to harvests caused by improved hunter access associated with logging road construction. Prior to 1965 most elk migrated through eastern portions of the herd unit whereas since 1965 a majority of elk migrate through Grand Teton National Park.

Table 2.6. *Number of elk crossing road transects in Grand Teton National Park. See Figure 2.4.*

Year	I. Gros Ventre Junction to Airport	II. Airport to Moose	III. Moose to Antelope Flats	IV. Deadman to Triangle X
1976	50	201	45	270
1977	0	46	64	70
1978	27	588	136	275
1979	63	585	206	78
1980	38	552	153	95
1981	31	111	60	65
1982	33	435	306	113
1983	11	148	109	140
1984	30	262	3	177
1985	36	466	208	96

Figure 2.12 Togwotee Pass area. Major migrations through this area have curtailed since c. 1965. Early migrations and open habitats render elk in this area particularly susceptible to hunter harvest. Photo by M. S. Boyce.

Chapter 3.

The elk population

Managing any wildlife species requires sound data on the size and composition of the population. Unfortunately, it is notoriously difficult to estimate the size of elk populations (Peek, 1985), although concentrations of elk on feedgrounds offer opportunity to count a fraction of the population. Another difficulty is that observations on the age and sex composition of elk populations are often biased. Counts of elk prior to the hunt often include a higher fraction of bulls because of approaching rut when bulls are generally less wary. Post-season counts often miss bulls, however, because they are less likely to visit winter ranges or feedgrounds, often staying at higher elevations in areas of deeper snow (Peek and Lovaas, 1968; Taber, Raedeke & McCaughran, 1982).

Despite various sampling difficulties, extensive counts and age/sex composition statistics have been compiled for the Jackson elk herd. These show pronounced patterns, some of which are surely independent of such sampling problems. In addition, reliable estimates of hunter harvest are obtainable through check stations and mail questionnaire surveys. My objective in this chapter is to review methodologies for estimating population parameters for elk, and to summarize trends in available data for the Jackson elk herd. Finally, I attempt to characterize mechanisms of population regulation for the elk herd through density-dependent birth and death processes.

Between 1912 and 1932, efforts to count elk in Jackson Hole during winter employed field crews of 10–12 men on skis, snowshoes or horseback. To minimize the possibility of duplicating counts, they attempted to count an entire drainage each day. Each annual winter census took approximately one month to complete. Only elk that were seen were recorded and no attempt was made to estimate the number of elk not counted (Anderson, 1958).

Beginning in 1932, aerial surveys were employed to ensure better coverage of elk winter ranges. Anderson (1958) details the methodology employed during these early aerial surveys. He also notes serious sampling difficulties with aerial surveys that make these counts unreliable, and not comparable from year to year. There appears to be a concensus that aerial survey techniques cannot provide a reliable estimate of the size of elk populations (Buechner, Buss & Bryan, 1951; Peek, 1985).

Recent research by Samuel *et al.* (1987), however, may offer hope for ranges on which numerous flights can be conducted each season. Samuel *et al.* calculated a sightability index from a ratio of radio-collared elk observed versus those known to be present but not seen. This technique allows an evaluation of various factors influencing sightability and thereby permits an estimate of population size. Currently, however, techniques are not adequately refined to justify extensive aerial counting of elk, particularly because airplane or helicopter rental can be expensive.

Counts on feedgrounds

Large concentrations of elk on feedgrounds offer a relatively easy opportunity to enumerate elk. Since feedgrounds were first established in Jackson Hole, counts have been conducted by observers riding on hay sleds. Standardization of methodology is extremely important to ensure comparability in counts from year to year. On the National Elk Refuge, a composition count where elk are classified according to sex and age is a coordinated effort usually conducted during a single day by a team of Wyoming Game and Fish Department, National Elk Refuge, Forest Service and National Park personnel. Timing of the count is chosen to coincide with the period of maximum concentration of elk on the feedground. This is assessed by the Refuge biologist who frequently surveys distribution and age/sex composition of elk on the Refuge.

Composition counts on the Gros Ventre feedgrounds are coordinated by the Wyoming Game and Fish area biologist. Counts are conducted from horse-drawn sleighs while the elk are being fed. Teams of 4–7 people classify elk on each of the three feedgrounds simultaneously so that no interchange between feedgrounds can confound counts. If not all elk can be counted in one day, the count is discarded and an additional count made on another day. Age and sex classifications of the elk are typically repeated 2–3 times each year with the best time usually being in late February when the greatest numbers of elk are attending the feedgrounds. Weather permitting, aerial surveys are conducted on the morning following a classification count to attempt an estimate of the number of elk wintering off feed on the Gros Ventre winter range.

In Figures 3.1 and 3.2, I present counts of the number of elk at the National Elk Refuge and the Gros Ventre feedgrounds (data in Appendices A and B).

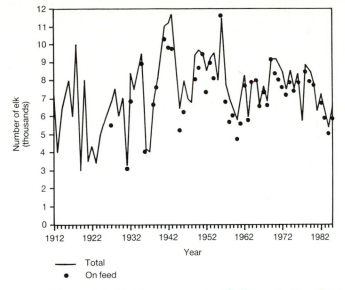

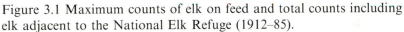

Total
• On feed

Figure 3.1 Maximum counts of elk on feed and total counts including elk adjacent to the National Elk Refuge (1912–85).

There have been no substantial long-term declines in the number of elk, contrary to some reports (Cole, 1969; Boyd, 1978). In fact, the total counts of elk on and adjacent to the National Elk Refuge show a statistically significant positive trend since 1912 ($r_s = 0.272$, $N = 57$, $P < 0.05$). The number of elk actually being fed at the National Elk Refuge shows virtually no change over a shorter run of years for which data are available ($r_s = -0.091$, $N = 45$, $P > 0.1$), but counts have declined significantly over the past ten years. Feeding on a

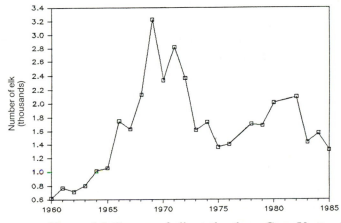

Figure 3.2 Counts of elk at the three Gros Ventre feedgrounds, 1960–85. No counts were made during the mild winters of 1977 and 1981 when no feeding took place.

regular basis began about 1960 at the Gros Ventre feedgrounds. The increase in subsequent years reflected an increase in the Gros Ventre winter population as well as an increase in the number of Gros Ventre elk habituated to feedgrounds.

Effects of climatic variation

Feedground counts do not offer a totally reliable index of population trends because weather conditions influence elk distribution (Anderson, 1958). Indeed, at least 50% of the variance in counts on the National Elk Refuge is attributable to winter severity (Sauer & Boyce, 1979b). During mild winters, a greater fraction of the elk herd winters away from feedgrounds, while during severe winters, more elk are likely to forage on feedgrounds.

Several variables interact to create 'winter severity', including temperature, snow depth, wind, snow packing and crusting (Murie, 1951c; Anderson, 1958; Skovlin, 1982). Snow depths in excess of 46 cm have been observed to restrict elk movements and influence distribution (Beall, 1974; Leege & Hickey, 1977; Sweeney & Steinhoff, 1976). Wilson (1958) claims that 'Snow depth apparently influences winter distribution [of the Jackson elk herd] much less than snow condition. The fact that the snow remained uncrusted until bare ground began to be exposed in March [1958] allowed the animals to dig through considerable depths in search of feed.' Wilson also notes that in 1958 approximately 8100 elk wintered on feedgrounds compared with nearly 15000 during 1955–56 when 'winter was characterized by extreme icing and crusted conditions of heavy snowfall interspersed with rain and thawing early in [the] season'.

Details on the condition of snow in Jackson Hole are not available, thus available data including maximum depth of snow on the ground for each month, monthly mean temperature, and total monthly precipitation must be relied upon to evaluate the effect of climatic variation on counts of elk on feedgrounds. Although several variables contribute to winter severity, January precipitation appears to be the best predictor of the maximum counts of elk on the National Elk Refuge. Consistently, over various collections of years, January precipitation is the winter climate variable which is most highly correlated with maximum counts of elk. Additionally, regression models of elk counts as a function of January precipitation are very stable; for example, a virtually identical model is obtained over the years 1960–85 as for 1936–85.

The relationship between the number of elk on feed at the National Elk Refuge (FEED), and January precipitation at Moose, Wyoming (PCP_{Jan}), can be expressed as a linear regression model:

$$FEED = 5529.6 + 24.3(PCP_{Jan}) + \epsilon$$
$$r = 0.7, P < 0.001 \tag{3.1}$$

A multiple regression model including both snow depth and temperature accounted for over 50% of the variance in the Refuge census over the period 1950–78, but when earlier years were included in the analysis, temperature did not contribute significantly to the model (Sauer & Boyce, 1979b). In addition, temperature was positively correlated with Refuge counts suggesting that warm winters with heavy snowfall resulted in the greatest attendance by elk at the National Elk Refuge. Updating this analysis to 1985, I found the opposite pattern, that is, more elk appear on the Refuge during cold winters with heavy precipitation. Therefore, temperature is not a reliable predictor of elk attendance at the Refuge.

Snow depth is highly correlated with elk counts on the Refuge, but January precipitation is more highly correlated and yields a more stable model irrespective of choice of weather station. January precipitation may be a better predictor than snow depth because precipitation not only reflects snow depth but also confounding effects of other forms of precipitation such as rain and sleet. A January thaw and rain followed by cold temperatures can result in extremely tough foraging for elk (Wilson, 1958).

Sex and age composition of elk on feedgrounds is also highly correlated with winter severity. In particular, mature bulls are more likely to winter away from feedgrounds if snow conditions permit (Boyd, 1978; Peek & Lovaas, 1968). Boyd (1978) notes that bulls may winter on ranges with up to 1.3 m of snow on the ground, and Anderson (1958: 140) remarks that 'larger bulls commonly winter in the higher, rougher portions of the range'. This accounts for a very strong correlation between the number of mature bulls per 100 cows on the Refuge and maximum January snow depth at Jackson ($r = 0.801$, $n = 14$, $P < 0.01$) and a similar relationship for the pooled composition data for the Refuge and Gros Ventre feedgrounds ($r = 0.816$, $n = 14$, $P < 0.01$).

Digestive efficiency for elk calves is reduced at low ambient temperatures (Westra & Hudson, 1981). This suggests the hypothesis that low temperatures might increase attendance of calves at feedgrounds. However, I found no support for this hypothesis; temperature during winter months and the proportion of calves on either the Refuge or Gros Ventre feedgrounds were not correlated ($P > 0.1$). Furthermore, there were no consistent patterns in the correlation coefficients, that is, the correlations were positive some months and negative other months. Possible reasons that calf visitation at feedgrounds is independent of temperature include (1) calf movements at eight months of age are still tied to those of the dam, and (2) the range of temperatures for feeding trials of Westra & Hudson (1981) was much greater than interannual variation in winter temperature in Jackson Hole.

These observations have important ramifications, because over-reliance on feedground counts can underestimate the elk population during mild winters. A linear modelling approach allows us to correct feedground counts for winter severity, thereby standardizing counts for a 'tough winter'. Determining conditions that constitute a 'tough winter', however, is necessarily subjective. Arbitrarily, I define a severe winter as being one in which January precipitation falls at the 90th percentile or greater, which we shall term P_{90}. For weather data from Moose, Wyoming, this defines severe winters as those with total January precipitation of 120.9 mm of precipitation or greater.

The estimated number of elk, $\hat{N}$, that would feed at the National Elk Refuge given a severe winter can be predicted with the following model:

$$\hat{N} = \text{FEED} + \beta_1(P_{90} - P_{\text{obs}}) \tag{3.2}$$

where FEED is the maximum number of elk observed on feed at the National Elk Refuge during a winter, P_{obs} is the total January precipitation observed during that year, and $\beta_1 = 22.28$. These 'corrected' population levels will have confidence intervals defined by

$$\hat{N} \pm 1.96\{\text{MSE}[1 + (P_{90} - P_{\text{obs}})^2/(n-1)\text{Var}(P)]\}^{\frac{1}{2}} \tag{3.3}$$

For severe winters, that is, those with more than 120.9 mm of January precipitation, $\hat{N} = \text{FEED}$. A confidence interval cannot project a smaller value than FEED. For these data, the mean square error (MSE) equals 713 069, and the variance in January precipitation (Var[P]) is 1694 ($n = 50$). I did not calculate a model to correct counts on the Gros Ventre feedgrounds because suitable weather data are not available for the Gros Ventre valley.

At Figure 3.3 and Table 3.1 I present original maximum counts of elk on the National Elk Refuge together with corrected estimates and confidence intervals

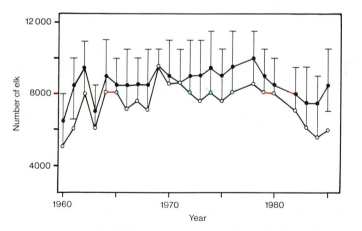

Figure 3.3 Corrected counts of elk on the National Elk Refuge, 1960–85, standardized for a winter of high January precipitation (pcp > 120.9 mm at Jackson). Solid dots are corrected counts, open dots are original counts, and error bars are 95% confidence intervals for the corrected estimates.

adjusting counts from mild winters upwards according to the model at equation 3.1. The consequence of adjusting counts to a standardized severe winter is to increase the mean count and to decrease the relative variability in the counts. Thus, average FEED = 7110 with CV = 15.7%, whereas average $\hat{N}$ = 8325 with CV = 10.0%.

These corrected feedground counts offer a better index to relative population size than original counts not corrected for January precipitation. However, a

Table 3.1. *Number of elk on the National Elk Refuge corrected for a tough winter. Corrected counts and confidence intervals were calculated with equations 3.2 and 3.3.*

Year	Number of elk on feed	Corrected counts	95% CI	
			Low	High
1960	4746	6259	4746	7959
1961	5591	8037	6267	9808
1962	7666	9268	7666	10973
1963	5827	6807	5827	8481
1964	7916	8974	7916	10651
1965	7946	8144	7946	9800
1966	6556	8374	6655	10094
1967	7369	8049	7369	9713
1968	6659	8426	6710	10142
1969	9205	9205	9205	10868
1970	8421	8920	8421	10580
1971	8054	8364	8054	10021
1972	7615	8901	7615	10588
1973	7194	8979	7261	10696
1974	7878	9458	7878	11162
1975	7450	8537	7450	10215
1976	7858	9382	7858	11082
1977		——no feeding——		
1978	8413	9460	8495	11137
1979	7828	8621	7828	10289
1980	7749	8230	7749	9890
1981		——no feeding——		
1982	6530	7363	6530	9032
1983	5878	7453	5878	9157
1984	5010	7189	5442	8936
1985	5758	8215	6444	9987
1986	6430	7506	6430	9184

better index of winter severity probably exists, which should be determined by future research. An improved index of winter severity will probably include more detail on characteristics of snow, for example, crusting and packing. However, characterization of snow structure solely at the National Elk Refuge may not be adequate, since snow conditions at higher elevations will determine whether elk are forced to migrate to lower elevations.

Central valley counts

The expansion of Grand Teton National Park in 1950 was accompanied by the immediate termination of elk hunting west of the Snake River (except the Moose Creek–Berry Creek area) and, after 1957, by elimination of all livestock grazing west of the Snake River (Righter, 1982). Prior to 1950, few elk summered in Grand Teton National Park (Cole, 1969). Apparently due to protection from livestock grazing and hunting (Martinka, 1969), numbers of elk summering in the Park increased dramatically. By 1958, substantial numbers of elk summered in central portions of the valley area of the Park (McLaren, pers. comm., as cited by Martinka, 1965: 21). In October 1962, an aerial count located over 2500 elk migrating from Grand Teton National Park onto the National Elk Refuge (Cole, 1969). By 1963, elk in the central valley comprised a substantial fraction of the National Elk Refuge winter herd (Martinka, 1969). Beginning in summer 1963, Martinka (1965, 1969) began research to estimate the numbers of elk using central valley portions of Grand Teton National Park.

Martinka (1965, 1969) developed a method for estimating the number of elk in the central valley portions of Grand Teton National Park based upon counts of elk along prescribed observation routes in the valley. Population estimates are derived from mark-recapture studies conducted in the early 1960s. A detailed description and evaluation of the technique is presented in Appendix B.

Precision of Martinka's central valley estimator seems reasonable (see Table 3.2), and several correlations with independent population data tends to support the validity of the estimator. Estimates are probably biased underestimates of the size of the central valley population, particularly when mature bulls are included in the estimates. Estimates of the cow-calf-yearling male segment of the central valley population are significantly correlated with counts of elk on the National Elk Refuge during the following winter, but there is no correlation between total valley estimates and Refuge counts. Again, this may be attributable to bias in estimating the population of mature bulls in the valley during summer.

Estimated number of elk summering in central valley areas of Grand Teton National Park, based upon summer valley counts, trends upward between 1963 and 1985 (Figure 3.4); however, this trend is not statistically significant ($r_s = 0.432, n = 22, 0.05 < P < 0.1$). Eliminating the low counts for 1965 and 1966 because elk were disturbed by a mountain pine beetle control operation (Cole, 1969) even further reduces the upward trend in estimated population size ($r_s = 0.31, n = 20, P > 0.1$). Estimated numbers of elk summering in central valley areas of the Park comprised from 11.5% of the Refuge winter herd (Figure 3.3) in 1976 to 23.9% in 1981 (Figure 3.5), with no detectable temporal trend ($r_s = 0.383, n = 18, P > 0.1$).

In contrast to these trends, the cow-calf-yearling male segment of the central

Table 3.2. *Summer trend counts, Grand Teton National Park*

Year	GTNP valley count	Valley population estimate	95% confidence		Willow Flats	Summer elk on National Elk Refuge
			Lower limit	Upper limit		
1963	732	1627	1190	2064	53	386
1964	687	1527	1117	1937	43	352
1965	371	824	610	1039	—	119
1966	401	891	658	1124	—	101
1967	513	1140	838	1442	—	223
1968	595	1322	970	1675	—	216
1969	644	1431	1048	1814	0	350
1970	780	1733	1266	2200	0	275
1971	509	1131	832	1431	8	80
1972	725	1611	1178	2044	0	42
1973	740	1644	1202	2087	0	12
1974	896	1991	1453	2530	0	47
1975	793	1762	1287	2237	11	7
1976	484	1076	791	1360	9	20
1977	649	1442	1056	1828	7	0
1978	810	1800	1315	2285	19	137
1979	750	1667	1218	2115	0	147
1980	701	1558	1140	1976	0	57
1981	628	1396	1022	1769	0	0
1982	789	1753	1281	2226	0	86
1983	—	—	—	—	—	150
1984	654	1453	1064	1842	0	100
1985	812	1804	1317	2291	33	47

valley population exhibits a declining trend over the period 1969–85 ($r_s = -0.482, n = 15, 0.05 < P < 0.1$; Figure 3.6). This appears to be attributable to the herd reduction program initiated in 1975 which focused on antlerless harvest of elk, and indeed, during years of the herd reduction program estimates of the cow-calf-yearling male segment of the central valley population are significantly lower than prior to 1975 ($P < 0.05$).

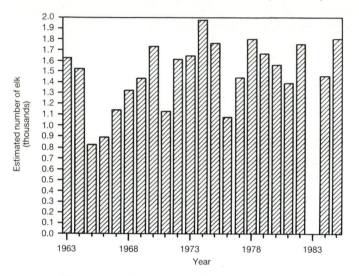

Figure 3.4 Estimates of the total number of elk in valley areas of Grand Teton National Park, 1963–85.

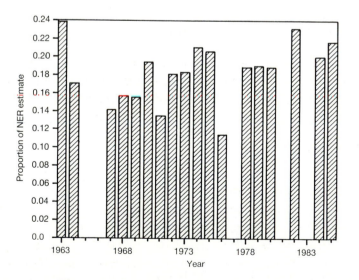

Figure 3.5 Estimates of the valley herd in GTNP as a fraction of the number of elk on the Refuge (corrected for a 'tough' winter).

Based upon earlier records (Murie, 1951c; Anderson, 1958; Cole, 1969), there is little question that the number of elk summering in Grand Teton National Park increased substantially subsequent to expansion of the Park in 1950 and after removal of livestock grazing in 1957. This conclusion is supported by track counts across Park transects, particularly on Burned Ridge (Figure 2.9). However, some of the elk crossing the Burned Ridge transect are likely to be coming from the Teton Range segments (e.g., Berry Creek, Owl Creek and Moose Basin), and north of the Park, in addition to elk summering in the central valley.

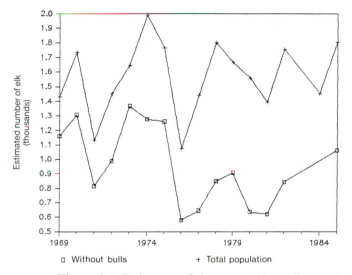

Figure 3.6 Estimates of the cow-calf-yearling male segment of the central valley elk population in Grand Teton National Park, 1969–85.

The central valley trend counts suggest little increase in the number of elk summering in the Park since 1963. However, due to differential visibility of mature bulls and the increased number of mature bulls in recent years, the total Park population may have increased more since the mid-1960s than the trend counts indicate.

Population structure

Understanding population structure can be important for management because ultimately it is the combination of birth and death rates that determine population growth rates. Ecological variables cause variation in birth and death rates which ultimately determine allowable harvest. In addition, there is interest in trophy hunting for mature bulls and therefore maintenance of a huntable population of mature bulls can be a priority for management by at least one interest group.

There are several sources of information on the dynamics of sex and age composition for the Jackson elk herd. The most extensive information includes composition counts on feedgrounds in winter, summer composition counts (mostly from Grand Teton National Park; see Appendix H), and composition of hunter harvest. In addition, over 11000 elk ear-tagged since 1943 (Allred, 1950; Anderson, 1958) afford estimates of survivorship (Straley, 1968; Sauer & Boyce, 1983).

In this section I review the demography of the Jackson elk herd with an aim of better appreciating the mechanisms that regulate population size. Man has mediated natural processes through winter feeding. Also, with hunting, man has almost certainly replaced natural sources of mortality which regulated natural populations of elk. By definition, population regulation implies that certain demographic parameters are density dependent. Calf survival in the Jackson elk herd is density dependent (Sauer & Boyce, 1983), despite the fact that elk in Jackson Hole receive supplemental food in winter.

Natality

During winter of 1935–36, 541 elk were culled in Jackson Hole. Of these, 312 were cows of breeding age ($2\frac{1}{2}+$), and all but 8 were pregnant (Murie, 1951c: 271). This suggests a pregnancy rate of 97.4%. Somehow, other reviews of these same data derived 89.2% of 334 cows older than yearlings to be pregnant (Anderson, 1958; Murie, 1951c: 142; Cole, 1969).

Cole (1969) examined reproductive tracts of 751 female elk killed on or adjacent to Grand Teton National Park between 1 October and 30 November from 1962 to 1966. Of cows older than calves, 78% ($n = 728$) were pregnant or possessed ovarian structures indicating fertility. This rate was 89% ($n = 620$) for cows older than yearlings, and 15% for yearlings ($n = 108$). Although Cole notes that some of these females may have been killed before the end of the breeding season, most should have conceived (Morrison, Trainer & Wright, 1959).

More recently, in conjunction with a feeding study conducted 1976–82 on the National Elk Refuge, a select group of 318 cows age $2\frac{1}{2}$ and older were palpated (Greer & Hawkins, 1967) to assess pregnancies. Here, 277 were pregnant, or an 87.1% pregnancy rate. Among yearling cows, only 17% ($n = 50$) were pregnant.

These pregnancy rates compare favorably with rates from other elk populations reviewed by Taber *et al.* (1982). Yearling pregnancy rates averaged 20% (range 0–48%) from 13 different reports. Pregnancy rates among 3.5–7.5 year old elk averaged 87% (range 49–99%) for eight populations.

Pregnant elk may lose their calves due to poor nutrition of cows (Thorne,

Dean & Hepworth, 1976), and after becoming infected with brucellosis, females often abort their first calves (Thorne, Morton & Ray, 1979). Pregnancy rates may decline with increasing density (Gross, 1969), particularly for yearlings (Houston, 1982). Pregnancy rates in the Northern Yellowstone herd seemed little affected by density during 1935–70 (Houston, 1982), consistently averaging about 82% when including yearlings and 87% for age 2 + cows. Since then, however, there appears to be a reduction in pregnancy rates at very high densities with only 71% ($n=810$) pregnant when including yearlings and 76% ($n=721$) for age 2 + cows in 1984 (J. Swenson, Montana Department of Fish, Wildlife & Parks, unpublished).

During feeding trials at the National Elk Refuge, Oldemeyer, Robbins & Smith (unpublished) held 93 pregnant cows which were trapped in January until parturition. Of these, 91.4% produced calves. Eight of ten *in utero* losses of calves were in brucellosis positive cows.

Calf sex ratio

From the cull of elk from Jackson Hole during the winter of 1935–36, Murie (1951c: 271) reports that pregnant females bore 144 male fetuses and 129 female fetuses (52.7% males). This is fairly typical of fetal sex ratios for elk, which have been observed in other studies to vary from 46.8% to 52.4% males, seldom showing statistically significant deviations from parity (Houston, 1982).

This tendency towards a 1:1 sex ratio among offspring is also reflected in the pooled overall sex ratio amongst calves killed by hunters. Between 1951 and 1985, 4547 male calves and 4459 female calves were killed in the Jackson herd unit. This reflects a sex ratio of 50.5% males which does not differ significantly from 1:1 ($\chi^2 < 1$, $P > 0.1$).

Closer inspection of the spatial and temporal pattern in the sex ratio of calves in the harvest reveals some marked patterns (Table 3.3). In particular, sex ratio of calves shot in Grand Teton National Park has declined from a majority of males prior to 1975 to a highly significant ($P < 0.01$) preponderance of females in more recent years (Figure 3.7). Correlated with this change in sex ratio has been a pronounced increase in the proportion of adult males in the Grand Teton National Park herd segment as a consequence of non-antlered elk harvest restrictions in the Park during years 1976–83 (Figure 3.8).

There are several possible explanations for the observed sex ratio among Park calves. First, female calves in the Park may be selectively taken by hunters. I find it difficult to imagine how hunters might be selective for the sex of calves, because the sex of calves can be difficult to assess in the field. Furthermore, if

Table 3.3. *Calf sex ratios from the Jackson elk herd. Sample size in parentheses.*

Year	Teton Wilderness	National Forest (South of Teton Wilderness area)	Grand Teton National Park and National Elk Refuge	Total Jackson herd
1951–54[a]				60.4% (599)**
1956				53.2% (502)
1957				50.0% (268)
1958				52.7% (393)
1959				54.4% (114)
1960				51.4% (288)
1961			36.8% (19)	36.8% (19)
1962			50.0% (36)	41.7% (120)
1963			45.9% (74)	53.2% (370)
1964			55.4% (130)	68.2% (352)**
1965	73.1% (52)**	63.7% (91)**	53.9% (89)	60.1% (393)**
1966	55.0% (40)	55.2% (67)	59.8% (122)*	57.6% (229)
1967			56.2% (73)	56.2% (73)
1968	57.5% (40)	56.5% (69)	51.3% (80)	54.5% (189)
1969	62.8% (43)	48.1% (131)	53.3% (90)	52.3% (264)
1970	43.5% (46)	52.6% (209)	47.7% (111)	50.0% (366)
1971	47.2% (72)	40.6% (143)*	53.8% (106)	46.4% (321)
1972	45.7% (35)	43.6% (94)	52.4% (84)	47.4% (213)
1973	50.0% (186)	45.5% (200)	45.5% (123)	47.2% (509)
1974			59.6% (47)	59.6% (47)
1975	62.5% (88)*	49.9% (389)	42.4% (144)	49.9% (621)
1976	77.8% (9)	59.9% (70)	51.6% (64)	53.8% (143)
1977	57.1% (28)	42.7% (157)	43.2% (125)	44.2% (310)
1978	57.1% (21)	45.0% (222)	40.2% (122)*	44.1% (365)*
1979	54.5% (33)	41.0% (295)**	37.3% (217)**	40.4% (595)**
1980	52.2% (23)	51.2% (41)	33.3% (75)**	41.7% (139)
1981	63.6% (11)	50.6% (237)	49.1% (175)	50.4% (423)
1982	48.1% (54)	53.7% (95)	39.9% (278)**	44.0% (427)*
1983	50.0% (36)	48.8% (41)	33.6% (217)**	37.8% (294)**
1984	— (2)	62.5% (16)	38.0% (50)	52.2% (69)
1985	—	—	30.4% (46)**	30.4% (46)**
1951–75	54.5% (602)*	49.6% (1393)	51.5% (1309)	53.4% (6245)**
1976–85	54.4% (217)	46.6% (1174)*	39.8% (1369)**	44.0% (2761)**
Totals	54.5% (819)*	48.2% (2567)	45.5% (2678)**	50.49% (9006)

*Deviation from parity, $P < 0.05$
**Deviation from parity, $P < 0.01$
[a] Anderson (1958: 139)

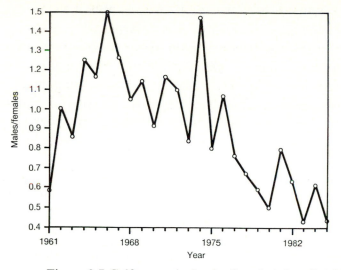

Figure 3.7 Calf sex ratio (males/females) for elk killed in Grand Teton National Park, 1961–85.

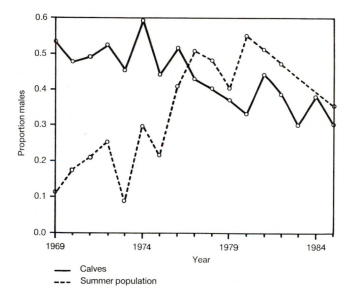

— Calves
- - - Summer population

Figure 3.8 Calf sex ratio (proportion males) and the proportion of adult males among summer composition counts in Grand Teton National Park, 1969–85.

hunter selectivity was affecting the sex ratio, why should the sex ratio be highly skewed for elk calves shot inside the Park whereas those shot outside the Park have a sex ratio near parity?

Another possibility is that cows in good condition should invest more in male offspring because male fitness can be more strongly influenced by parental investment (Trivers & Willard, 1973; Austad & Sunquist, 1986). Such a difference in parental investment may occur before or after birth, although since fetal sex ratios have never been shown to differ from parity, the observed patterns most likely occur as a consequence of differential mortality among male versus female calves. For red deer (also *Cervus elaphus*), indirect support for the Trivers & Willard hypothesis was offered by Clutton-Brock, Albon & Guinness (1984) who showed that dominant females were more likely to produce surviving male offspring, and that maternal rank had greater effect on the breeding success of sons than of daughters.

Indeed, one might argue that condition of elk summering in Grand Teton National Park has declined in recent years due to sustained high elk densities in the Park since the mid-1960s, and therefore cows may be investing less in male calves. This may be supported by an inverse correlation between the sex ratio of calves in the Park harvest and estimates of the total central valley population in summer ($r = -0.454$, $n = 22$, $P < 0.05$). However, this is inconsistent with the fact that harvest pressure on Park elk has been particularly high starting in 1976 with the implementation of antlerless hunting restrictions in the Park, resulting in decreased estimates of the cow-calf-yearling male segment of the central valley population. In fact, increasing recruitment rates suggest that, if anything, range condition in the Park has improved (see Figure 3.9).

An alternative explanation is that male calves may be more susceptible to food shortage due to their faster growth rates and higher nutritional require-

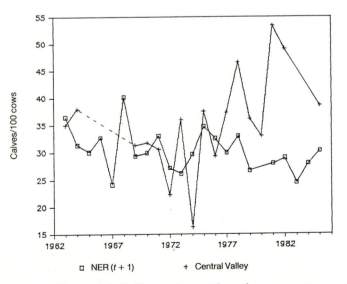

Figure 3.9 Calf recruitment based upon summer composition counts in GTNP and in the following winter composition counts at the National Elk Refuge, 1963–86.

ments (Clutton-Brock, Albon & Guinness, 1985). High-density populations in the Park may have reduced conditions which are conducive for calf survival in the Park, and consequently, impacted males more than females. Again, there is no evidence for this because calf recruitment in the Park increased between 1969 and 1985 based upon valley composition counts ($r_s = 0.527$, $n = 17$, $P < 0.05$), and recruitment in the Park is certainly no lower than average among elk wintering on the National Elk Refuge (Figure 3.9). Also, if the Clutton-Brock, Albon & Guinness (1985) hypothesis were true, precipitation should be positively correlated with sex ratio, because rainfall can improve range condition (Picton, 1984). This is true for precipitation in June ($r = 0.432$, $n = 23$, $P < 0.05$); however, this observation seems equivocal since July precipitation is inversely correlated with sex ratio ($r = 0.499$, $n = 23$, $P < 0.05$). Whereas the Clutton-Brock *et al.* hypothesis may well explain why young males generally suffer higher mortality among polygynous species, it does not appear to account for the patterns of sex ratio variation in the Jackson elk herd.

Several species of cervids have been shown to have male-biased sex ratios when nutrition of the dam was poor (Verme, 1983; Skogland, 1986). Verme's (1983) hypothesis is that male calves are more likely than female calves to disperse out of local areas of poor resource abundance. Thereby, females in poor habitat patches may enjoy higher fitness if they bear male calves or if they invest more heavily in their male offspring. It is possible that coyote (*Canis latrans*) predation is higher on male than female calves because male calves are more rambunctious (Klein, 1970). Certainly coyote populations are high in Grand Teton National Park (Weaver, 1979a; Camenzind, 1978) and cow elk defend their calves very vigorously against coyotes (Altmann, 1952). However, substantial predation on elk calves by coyotes has not been reported, and I am unaware of any evidence suggesting that coyote populations have increased during the past few years.

Finally, the inverse correlation between calf sex ratio and the proportion of bulls in the valley ($r = -0.563$, $n = 15$, $P < 0.05$, Figure 3.8) may be artifactual, or it may suggest local resource competition among males. Competition among female red deer depressed the reproductive success of adult females, offering an advantage to producing male young (Clutton-Brock, Albon & Guinness, 1982). This explanation seems less plausible for males, however, because in polygynous species, males generally have a greater propensity to disperse (Dobson, 1982). However, adult male elk in Grand Teton National Park show a high propensity to return to the same area each summer (B. Smith, unpublished).

Local resource competition among males may be the most parsimonious explanation for the patterns in calf sex ratio in the Park. Yet, because male elk

Table 3.4. *Recruitment rates for Jackson elk herd assessed by ratios of number of calves per 100 cows in composition counts and harvest classifications.*

	(1) NER winter classification counts	(2) Gros Ventre winter classification counts	(3) Pooled (1) and (2)	(4) GTNP summer classification counts	(5) Harvest from GTNP	(6) Harvest from total Jackson herd unit
Average calves/100 cows	30.7	31.1	30.8	35.0	33.5	29.8
Standard error	1.19	1.32	1.10	2.66	1.40	1.29
Minimum	18.6	15.7	18.2	16.0	17.8	14.7
Maximum	44.7	45.7	44.8	53.0	47.9	41.8
Number of years	25	24	24	14	24	25

do not offer paternal care, any parental manipulation of sex ratio is most likely to come from the cow. Other examples of local resource competition appear to result from competition among females, for example, in galagos (*Galago crassicaudatus*; Clark, 1978) or the red deer studies of Clutton-Brock, Albon & Guinness (1982), therefore these data do not fit patterns offered for other mammalian sex ratio disparities involving local resource competition.

Reproductive opportunities for sons are almost certainly lower than for daughters in a population so overburdened with excess males. But, sex differential in adult mortality has no consequence to the optimal investment in offspring (Leigh, 1970). Yet in a long-lived species like elk, it is possible that bearing philopatric sons may reduce reproductive opportunities for other sons. Or perhaps male calves are more likely than female calves to be injured by aggressive bulls during the high-intensity rut reinforced by the high bull/cow ratios in the Park, thereby accounting for the inverse correlation between calf sex ratio and the proportion of bulls in the Park population. Murie (1951c) reports calves being impaled by antlers of bulls.

The manipulation afforded by the antlerless culls in Grand Teton National Park between 1976 and 1982 created some fascinating sex ratio changes, and almost certainly some marked consequences to elk social behavior as well. It is unfortunate that no one had the foresight to research the consequences of this manipulation.

Recruitment

Calf mortality is typically high during the first few days of life, thus pregnancy rates do not reflect recruitment into the population (Taber *et al.*, 1982; Greer, 1966). Available data allow three measures of recruitment into the Jackson elk herd: (1) calf/cow ratios from summer composition counts in central valley portions of Grand Teton National Park, (2) calf/cow ratios for elk wintering on the National Elk Refuge and Gros Ventre feedgrounds, and (3) calf/cow ratios of elk killed by hunters.

Recruitment reflected by these measures is similar, and has averaged 31.8 calves/100 cows overall since 1950 (Table 3.4). The calf/cow ratio from Grand Teton National Park in August averages 35 calves/100 cows which is significantly higher than 30.7 calves/100 cows observed on the National Elk Refuge during the following winter ($P < 0.005$). Also, kill within the Park yields a significantly higher calf/cow ratio (33.5/100) than for the herd unit overall (29.8/100; $P < 0.005$), as should be expected.

Recruitment reflected by valley counts has increased significantly over the past 14 years ($r_s = 0.527$, $n = 17$, $P < 0.05$), but no significant trends occur in any of the other data sets (Figures 3.9, 3.10, 3.11 and 3.12). Calf/cow ratios are

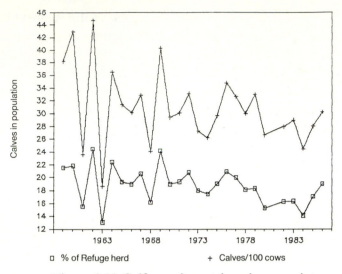

Figure 3.10 Calf recruitment based upon winter composition counts at the National Elk Refuge, 1959–86, expressed as a percentage of total Refuge herd and as calves/100 cows on the Refuge.

significantly correlated between Park summer composition counts and those taken the following winter on Gros Ventre feedgrounds ($r = 0.564$, $n = 14$, $P < 0.05$). Curiously, however, I found no correlation between composition counts of Park elk and those from the National Elk Refuge in the following winter ($r = 0.025$, $n = 13$, $P > 0.1$), nor between calf/cow ratios on Gros Ventre feedgrounds and those from the Refuge ($r = 0.335$, $n = 24$, $P > 0.1$). Calf harvest

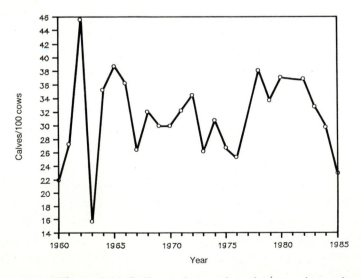

Figure 3.11 Calf recruitment based upon winter classifications on the Gros Ventre feedgrounds, 1960–85. Data from Appendix D.

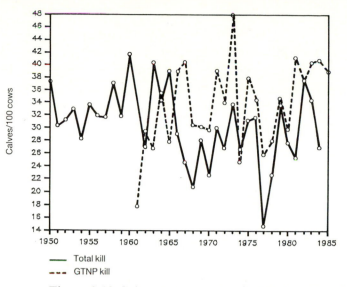

Figure 3.12 Calves/100 cows for elk killed in Grand Teton National Park and for elk killed in the entire Jackson herd unit, 1950–85.

appears to sample the composition of calves in the population as reflected by the positive correlation between the proportion of calves in the harvest of elk from Grand Teton National Park and the calf/cow ratios from composition counts in the Park the preceeding summer ($r = 0.601$, $n = 14$, $P < 0.05$).

Anderson (1958: 145) expressed considerable concern that recruitment rates were low for the Jackson herd. Indeed, during the period of Anderson's (1958) work, calf/cow ratios were often lower than during the past 30 years. Low recruitment rates in early years may have been a consequence of inadequate winter feeding rations (see Chapter 5), which is known to reduce survival of elk calves (Thorne, Dean & Hepworth, 1976). Average winter classification of approximately 30 calves/100 cows for the Jackson herd is not particularly low compared to other populations in the western United States (Taber *et al.*, 1982), especially since recruitment in the Jackson herd is reduced 8–12% by brucellosis infection (Thorne *et al.*, 1979; Oldemeyer, Robbins & Smith, unpublished).

Consider calf/cow ratios in view of no contribution expected from yearling cows and only 17% contribution to fecundity from two-year-old cows. By eliminating yearlings from the denominator and only counting 17% of an average 24% two-year-olds estimated from ear-tag returns by Straley (1968), we find that recruitment amongst mature cows now averages approximately 66 calves/100 cows. Given pregnancy rates of approximately 90%, this indicates that less than a third of the pregnant cows in the herd lose their calves by the time of winter composition counts.

Survival

Although undeniably a key in any population study, reliable estimates of survival are often missing. There are several reasons for this
(Caughley, 1977), not the least being that sample sizes required to achieve
reasonable accuracy are very large. For example, to establish a ±5% confidence interval on a survival estimate (or any proportion) demands a sample of
over 400 (Sokal and Rohlf, 1981). Yet, a 5% difference in survival for a long-
lived species such as elk can have serious consequences to population dynamics
(Nelson & Peek, 1982; Eberhardt, 1985).

Even though large sample sizes of known-age tagged elk exist from the
Jackson elk herd, confidence intervals are still larger than we would like for
reconstructing details of demography through time. In this section I summarize
what is known about survival patterns for elk in the Jackson elk herd based on
tagging studies as well as extrapolation from age and sex classifications.

Tagging studies

Over 11 000 elk have been ear-tagged in Jackson Hole since 1943. Straley (1968)
analyzed returns from these ear tags using the composite-dynamic method to
estimate average survival rates. This method was subsequently discredited
because it unrealistically assumes constant survival over time (Burnham &
Anderson, 1979). However, models for analyzing tag recovery data have been
developed (Brownie *et al.*, 1985).

Data structure typical of tag recovery models is presented in Table 3.5. For
the specific case illustrated in Table 3.5, only adults are tagged and all adults are
assumed to have the same survival rates. Other models presented by Brownie *et
al.* (1985) relax these assumptions. Choice of the best model is determined by a
chi-squared goodness of fit test. Typically, the larger the data set, the more
sophisticated the model that will be selected. For our analysis, we have eliminated years with small samples because spurious estimates can obtain.

From program BROWNIE (Brownie *et al.*, 1985) we calculated survival estimates for elk tagged in the Jackson herd unit, mostly on the National Elk
Refuge (Sauer & Boyce, 1983). Model selection procedure follows Occam's
razor (Hutchinson, 1978) to use the simplest acceptable model. Alternative
models are used if and only if the data are adequate to justify the more complex
model. This approach ensures the smallest confidence intervals because the
degrees of freedom are kept small. If a priori information ensures that a more
complex model is necessary, it should be used. The goodness of fit test employed
by program BROWNIE attempts to optimize accuracy and precision of
estimators.

Three models were used for estimating survival. For cows and female calves for 1958–60, the best fit was obtained for a model that assumes variable survival and recovery rates from year to year, and also that first-year recovery rates differ from those tagged previously (Model H_2). For cows and female calves tagged 1962–67, the best model has year-specific survival and recovery rates, but first-year recovery rates are identical to those for earlier tags (Model H_1). This same model also fits tag recoveries for male calves and bulls tagged 1951–54, whereas males tagged 1958–60 were best analyzed with Model H_{02} which has constant survival rates but year-specific recovery rates.

I suspect that model H_2 is actually the appropriate model for all groups, but sample sizes were not adequate to justify selection of the more complex model using the model selection criteria of Brownie *et al.* (1985). Because of reduced degrees of freedom associated with the more complex model, confidence intervals surrounding survival estimates would be even greater.

Yearly estimates of survival (Table 3.6) vary substantially from year to year, and even with large sample sizes (pooled $n = 4067$), standard errors are large. Estimates of cow survival are significantly lower ($P < 0.05$) than Straley's (1968) estimate of 0.742. As expected, bull survival is lower than cow survival, but standard errors are so large that differences are not quite significant ($0.05 < P < 0.1$).

The large standard errors surrounding these estimates of survival are discomforting, yet these are some of the best survivorship statistics for any large

Table 3.5. *General structure of expected tag recoveries for models developed to estimate survival rates. The example presented here is for survival which varies from year to year but is constant with age. More sophisticated estimators where these assumptions are relaxed are presented in Brownie et al. (1985). Proportion of an age-class surviving = $\hat{S}$, and proportion of tags returned = $\hat{f}$.*

Year tagged	No. tagged	Expected recoveries in year					Row total
		1	2	3	4	5	
1	N_1	$N_1 f_1$	$N_1 S_1 f_2$	$N_1 S_1 S_2 f_3$	$N_1 S_1 S_2 S_3 f_4$	$N_1 S_1 S_2 S_3 S_4 f_5$	R_1
2	N_2		$N_2 f_2$	$N_2 S_2 f_3$	$N_2 S_2 S_3 f_4$	$N_2 S_2 S_3 S_4 f_5$	R_2
3	N_3			$N_3 f_3$	$N_3 S_3 f_4$	$N_3 S_3 S_4 f_5$	R_3
$k = 4$	N_4				$N_4 f_4$	$N_4 S_4 f_5$	R_4
Column totals		C_1	C_2	C_3	C_4	C_5	

Define: $T_1 = R_1$; $T_i = R_i + T_{i-1} - C_{i-1}$ $(i = 2, \ldots, k)$; $T_{k+j} = T_{k+j-1} - C_{k+j-1}$

Estimators: $\hat{f}_i = R_i C_i / N_i T_i$ $i = 1, \ldots, k$

$\hat{S}_i = [R_i(T_i - C_i)/N_i T_i] [N_{i+1}/R_{i+1}]$ $i = 1, \ldots, k-1$

mammal population. Most attempts to construct life tables for large mammals, for example, for the northern Yellowstone elk herd (Houston, 1982), require assumptions of stable or stationary age distribution, which are invalid (Caughley, 1977). A stable age distribution cannot exist for a long-lived species which is fluctuating substantially (Menkens & Boyce, 1989), as is true for the Jackson elk herd (Sauer & Boyce, 1979b).

An independent, unbiased estimate of male calf survival (Figure 3.13) is possible from estimates of the number of calves and yearling males on the National Elk Refuge in sequential years, assuming that half of the calves are males. Dispersal of elk out of the Jackson herd unit is low as shown from analysis of ear-tag returns in Chapter 2, and we presume that it is insignificant. Estimated male calf survivorship is

$$p_{1t} = (X_{t+1} \cdot S_{t+1}/N_{t+1})/(0.5 \ X_t \cdot C_t/N_t) \tag{3.4}$$

where p_{1t} is the probability of survival from 8–10 months of age to 20–22 months, S_t is the number of spike bulls and C_t is the number of calves

Table 3.6. *Estimates of elk survival rates ($\hat{s}$) of elk tagged and recovered in northwestern Wyoming.*

Sex	Year	Adults $\hat{s}$	Adults SE	Calves $\hat{s}$	Calves SE
Female	1958	0.2926	0.0678	0.3250	0.2972
	1959	0.8879	0.1520	1.0000	0.1438
	Mean	0.5903	0.0671	0.6885	0.1651
		610:203[a]		567:225[a]	
	1962	0.4110	0.1054	0.5488	0.1062
	1963	0.7402	0.0877	0.6553	0.1317
	1964	0.7174	0.1349	0.6210	0.1306
	1965	0.9202	0.1926	0.5745	0.1194
	1966	0.6063	0.1265	0.6604	0.1696
	Mean	0.6790	0.0300	0.6120	0.0596
		766:302[a]		674:240[a]	
Male	1951	0.2626	0.1366	0.2944	0.0796
	1952	0.4925	0.1749	0.2261	0.0764
	Mean	0.3776	0.0950	0.2603	0.0552
		72:28[a]		521:144[a]	
	1958–60 Mean	0.5362	0.0512	0.6614	0.1367
		276:105[a]		579:244[a]	

[a] Tagged:recovered.

among N_t elk classified in year t, and X_t is the estimate of the population of elk at the National Elk Refuge, standardized for a tough winter (from Table 3.1).

Mean survival from 1960 through 1985 is 0.712 ± 0.033 (SE). For 1961, $p_1 = 1.028$, and for 1982, $p_1 = 1.032$, which are obviously absurd values because they indicate greater than 100% survivorship. This could have occurred for a variety of reasons affecting estimates of the number of spikes or the number of calves, but I know of no reason why these years should be unusual.

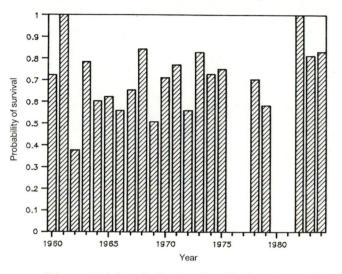

Figure 3.13 Survival of male calf elk estimated by dividing the estimated Refuge population of yearling males in year $t + 1$ by the estimated number of males calves in year t.

Age ratios

Despite the inability to estimate survivorship directly from age ratios, the data nevertheless can show patterns, which, when combined with other sources of information, may offer important insight into population mechanisms. For example, calf/cow ratios offer useful information on recruitment because we can usually assume very high survival of cows relative to that of calves.

Classification of elk also includes 'spikes' which are yearling males (Figure 3.14). Spike/cow ratios could provide information on survival of male calves to yearlings; assuming negligible mortality of cows. Unfortunately, spike/cow ratios can increase either because survival of calves is high, or because survival of cows is relatively low. Also higher dispersal rates for spikes than cows could confound spike/cow ratios.

Ratios of spikes/100 cows from summer composition counts in Grand Teton National Park when compared to the ratio of calves/200 cows (assuming 1:1 sex

Figure 3.14 Spike bull (yearling male) on the National Elk Refuge, 14 March 1986. Photo by M. S. Boyce.

ratio among calves) and male calves/100 cows (see Figure 3.7) show that spike/cow ratios cannot reflect survivorship of male calves because the ratios average higher than male calves/100 cows (Figure 3.15)! The valley spike/cow ratios vary much more than those from the National Elk Refuge and average much higher (Figure 3.16). The reason for the high proportion of spikes in the valley is because during migration more spikes remained there than expected (Martinka, 1969).

Spike/cow and calf/cow ratios on the National Elk Refuge seem more in line with each other (Figure 3.17). No trends are apparent over the past 30 years, although some of the earliest composition counts showed remarkably low spike/cow ratios (Figure 3.18).

Proportions of bulls on the National Elk Refuge (Figure 3.19) and in Grand Teton National Park during summer (Figure 3.20) show some exceptional trends, which are largely due to changes in hunting regulations (see below).

The fraction of bulls occurring on Gros Ventre feedgrounds shows no trends, and has not been so drastically influenced by changes in harvest regimen (Figure 3.21), except that the high proportion of spike bulls during 1972–75 can be attributed to spikes-excluded harvest regulations during those years. Gros Ventre counts show an average of only 2% bulls which is remarkably lower

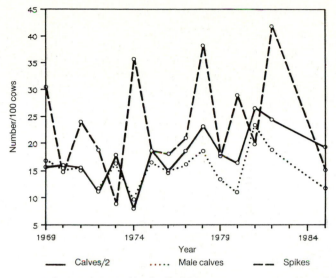

Figure 3.15 Spike bulls/100 cows among composition counts from summer counts in valley areas of Grand Teton National Park, compared with two estimates of the number of male calves/100 cows from the same composition counts. These include (1) half total calves per 100 cows, and (2) total calves times harvest sex ratio per 100 cows.

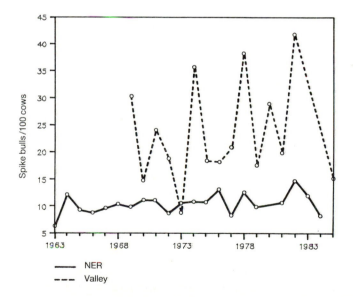

Figure 3.16 Spike bulls/100 cows in Grand Teton National Park during summer versus those on the National Elk Refuge the following winter.

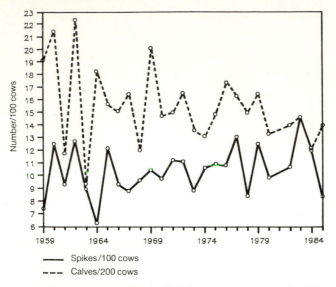

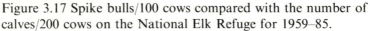

Figure 3.17 Spike bulls/100 cows compared with the number of calves/200 cows on the National Elk Refuge for 1959–85.

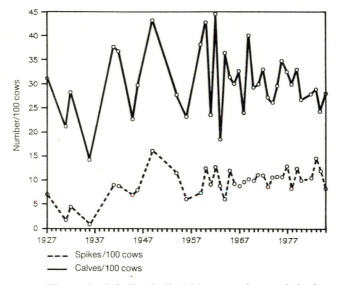

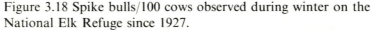

Figure 3.18 Spike bulls/100 cows observed during winter on the National Elk Refuge since 1927.

than for the National Elk Refuge. Again, this is attributable to differences in the harvest regimens experienced by different segments of the population.

Longevity

Bulls usually enter winter in poorer condition than females because of energy expended during the rut (Clutton-Brock, Albon & Guinness, 1982). Conse-

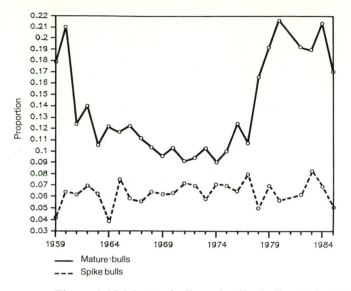

Figure 3.19 Mature bulls and spike bulls on the National Elk Refuge, 1959–85.

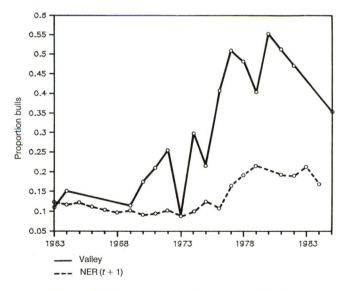

Figure 3.20 Proportion of mature bulls in summer classification counts in Grand Teton National Park, compared with the proportion of bulls appearing on the National Elk Refuge the following winter, 1963–85.

quently, overwinter mortality rates of males are typically double those for females (Straley, 1968; Smith, 1985).

Tagging studies (Straley, 1968) and age determinations by tooth sectioning (Keiss, 1969) provide the best age basis for longevity in the Jackson elk herd.

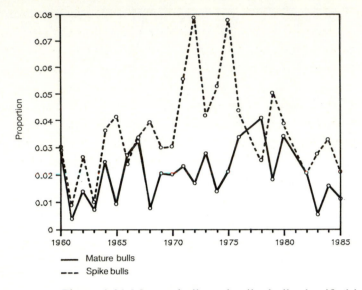

Figure 3.21 Mature bulls and spike bulls classified in winter on the Gros Ventre feedgrounds. Data from Appendix D.

Higher mortality rates for males (Sauer & Boyce, 1983) imply higher longevity for females. Only 3% of 2058 known-age, ear-tagged elk were older than 10 years, and most of these were females (Straley, 1968). Only two males survived to age 13, and all elk older than 13 were females. Two cows were older than 20. Three cows which died on the National Elk Refuge were older than 20 years (Smith, 1985).

Hunter harvest

Hunters account for most mortality in the Jackson elk herd. Harvests vary substantially from year to year depending upon factors such as harvest regulations, snow conditions, migration patterns, and hunter access into areas hosting substantial elk numbers. Hunting is clearly the most important tool available to managers for regulating the abundance and distribution of elk in Jackson Hole, and it provides extensive recreation and meat for thousands of hunters every year.

Design of hunt areas

Hunting is currently controlled regionally by specific regulations for 11 hunt areas (Figure 3.22). These hunt areas are not intended to reflect population units, but rather to permit improved spatial control and monitoring of harvest within the Jackson herd unit. Seasons and harvest quotas may differ in each hunt area, although regulations usually have been the same over blocks of these areas.

The current hunt area boundaries were established in 1971 and have been consistent since then. I have interviewed Francis Petera and James Yorgason who were the warden and biologist, respectively, when most of the current boundaries were established. Criteria for delineating these boundaries are described below.

Areas 70 and 71 were designed to divide the Teton Wilderness into units which comprised mostly Yellowstone National Park migratory elk (Area 71), and mostly Gros Ventre feedground elk (Area 70) so these herd segments could be hunted with different intensity. The northeast boundary of the herd unit (Area 60) is largely a jurisdictional boundary along the continental divide. Some elk that winter in the Jackson herd unit summer east of this boundary (Bendt, 1960). Most elk that summer east of the Yellowstone River generally

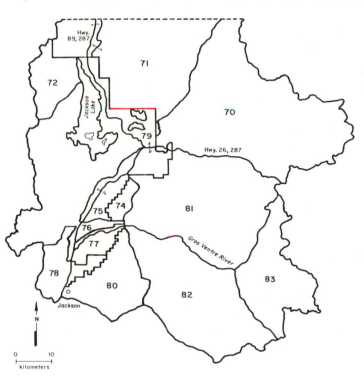

Figure 3.22 Hunt areas for the Jackson herd unit.

winter in drainages near Cody and Meteetse, Wyoming; whereas those that summer west of the Yellowstone River winter in the Jackson unit (Anderson, 1958; Cole, 1969).

In the Gros Ventre River drainage, Area 82 is maintained for mostly wilderness hunting with seasons and regulations usually similar to those for the Teton Wilderness. On the other hand, Areas 81 and 83 were established to allow better control in areas which were increasingly accessible from roads constructed to harvest timber and to drill exploratory oil and gas wells. Area 83 was the first in the Gros Ventre River drainage to receive heavy hunting pressure from four-wheel-drive vehicles gaining access along roads constructed in the Fish Creek vicinity. Substantial summering herds of elk occur in Area 81 along Cottonwood Creek, Charmichael Fork of Slate Creek, Moosehorn, Two Water and Bearpaw Creeks.

Hunt areas within Grand Teton National Park are natural units based upon areas which provide the greatest kill. Assigning hunters to specific areas within the Park ensures better distribution and lower density of hunters, thereby improving the quality of hunting. Area 74 was originally established because during migration of elk across the Park, a substantial migration route crossed the Snake River near Paintbrush Point and the Triangle X Ranch, then crossed Antelope Flats to Ditch Creek, where they were hunted. Low kills have been obtained from Area 74 since 1974 and since 1984 it has been combined with Area 81. Harvest statistics from the hunt areas are in Appendix E.

In addition to these 11 hunt areas, since 1963 hunters checking their kill at the Dubois check station must also report the location of their kill by drainage and special management area as mapped in Figure 3.23. Total kill from each of these 12 special management areas is summarized in Appendix F.

Management of the magnitude, composition, and spatial distribution of hunter harvest is accomplished by setting seasons, permit quotas, and harvest regulations. These regulations are administered by the Wyoming Game and Fish Department outside Grand Teton National Park, and jointly by the Game and Fish Department and the National Park Service for the hunt occurring within the boundaries of Grand Teton National Park.

Kill of elk from the Jackson herd unit is monitored primarily by (1) a mail survey of hunters which is contracted by the Wyoming Game and Fish Department to the Institute for Policy Research at the University of Wyoming, (2) check stations, and (3) field checks by the Wyoming Game and Fish Department. In addition, hunters must report all elk killed on the National Elk Refuge and within Grand Teton National Park to the Refuge and Park, respectively.

Mail surveys

Questionnaires (Figure 3.24) are mailed to hunters mostly selected from a computerized random draw of license holders. Another random sample is made for individuals whose license record is not computerized, for example, most resident general license holders who purchased their license over the counter. Hunters are sampled according to their license type (general, limited quota), and in the case of limited quota licenses, they are sampled by the assigned area of the limited quota license. Resident general license holders are

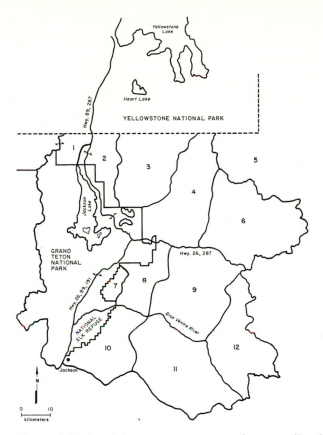

Figure 3.23 Special management areas for recording harvests checked at the Dubois check station.

sampled by proportional allocation by license sales in their county of residency. Sampling rates vary for different sampling frames, with slightly higher sampling intensity for limited-quota license holders than for general license holders. The questionnaire is designed to obtain information on hunter days, location, sex and age of the kill, but is kept simple to minimize response burden (Filion, 1981). A map showing all hunt areas in the state, along with current area regulations, is enclosed with each questionnaire to facilitate correct reporting of hunt areas.

In 1985, 28 777 questionnaires were mailed to elk hunters, which is 54% of approximately 53 000 elk license holders. Nonrespondents were mailed a second request, follow-up questionnaire. After these two mailings, a statewide response rate of 57% was obtained. Because of the uniqueness of license regulations for the Park and Refuge hunts, statistically larger sample sizes are drawn. Also, a special sample of hunters is drawn from the National Elk Refuge's list of permit holders after the season ends.

```
1985 Wyoming Elk Survey          No 33997
```

1. Please check Wyoming hunting licenses YOU held during 1985.

General Elk ☐   General Elk and Area 79 ☐   Limited Quota Elk ☐   Deer ☐

Antelope ☐   Small Game ☐   Bird ☐   Bear ☐   Big Game Archery ☐

2. Did you HUNT elk during the 1985 season?   Yes ☐ No ☐

3. If you were issued a limited quota elk license, what was the AREA(s) and TYPE?   Area Issued Type

4. For each AREA hunted using your general or limited quota license, indicate number of DAYS hunted and if you hunted the area on its OPENING DAY.

(use elk map)	1st area	2nd area	3rd area	4th area	5th area
Hunted AREA number					
DAYS hunted					
Hunted on area's OPENING DAY	Yes☐	Yes☐	Yes☐	Yes☐	Yes☐

5. Did you HARVEST an elk?   Yes ☐ No ☐

6. If you harvested, indicate the KIND, specific AREA of harvest and WEAPON used for harvesting.

Mature Bull☐   Spike Bull☐   Cow☐   Calf☐   Harvest Area

Weapon Used   Firearm ☐   Archery ☐

BEAR HUNTERS ONLY

7. Did you HUNT bear during the 1985 seasons?   Yes ☐ No ☐

8. Did you HARVEST a bear in 1985?   Yes ☐ No ☐

9. While hunting elk, were you contacted by Wyoming Game and Fish personnel?   Yes ☐ No ☐

Figure 3.24 Questionnaire used to sample elk hunters in mail surveys.

Since 1977, nonrespondents have been sampled each year by telephone, to determine differences between mail respondents and nonrespondents. Several types of response bias have been documented. First, there is a tendency for license holders who actually hunt to be more likely to return their mail questionnaires (hunter bias). Second, hunters successfully killing an elk return their mail questionnaires at a higher rate than unsuccessful hunters (success bias). Third, these two biases vary considerably depending upon license type (general versus limited quota), and residency (resident versus nonresident). A statewide comparison of the 1985 mail and telephone survey is shown in Table 3.7. Due to these biases, 1985 harvest estimates were reduced statewide by 3% for hunter numbers and 12% for total harvest (Cochran, 1977: Chapter 13).

It is sometimes impossible to determine bias by another survey because comparisons of surveys do not eliminate bias. For example, a telephone survey may also be biased in some way. Yet I think that we correctly understand the source of bias in the mail survey. For some telephone surveys, there is a substantial recall bias, for example, fishermen may not accurately recall exactly how many fish they caught. However, I believe that an elk hunter is unlikely to forget whether he or she hunted or killed an elk.

Estimates of hunter harvest have not always been so sophisticated. Prior to

1958, harvest was estimated from check station reports and estimates (guesses) by the district game warden. Beginning in 1958, the University of Wyoming Department of Statistics was contracted to estimate hunter harvest with mail surveys, but sampling intensity was considerably lower than it has been in recent years. For example, in 1959, the first year that a University of Wyoming IBM computer was used to analyze harvest reports, 3222 questionnaires were mailed with a 61.1% response rate (Wyoming Game and Fish Department, 1959). This compares with 28 777 questionnaires and a 57% response rate in 1985.

The Wyoming Game and Fish Department was concerned that the mail survey was not adequately monitoring the harvest, therefore in 1959 they launched an extensive (and expensive) 80% survey of hunters. Results from this survey differed from the 20% University of Wyoming survey by only 3%. Beginning in 1964, nonrespondents to the mail survey were sent a second follow-up questionnaire which increased the response rate.

The first attempt to measure nonresponse bias occurred in 1970 when 285 randomly selected resident nonrespondents were telephoned (2000 were telephoned in 1984). Nonhunters comprised 10.6% of the respondents to the mail surveys, but 25.2% amongst respondents to the follow-up telephone survey. In 1970, 37% of the mail survey respondents killed an elk compared to only 29.8% amongst nonrespondents interviewed in the follow-up telephone survey. Nonresponse bias is highest for hunters shooting a calf, and lowest for those killing a bull (Table 3.8).

Check stations

Although temporary check stations have been used from time to time to monitor hunter harvest coming from the Jackson elk herd, the permanent check station east of Dubois provides the only long-run data set for evaluating

Table 3.7. *Comparison of mail and telephone surveys of elk hunters in Wyoming during 1985.*

	General license holder		Limited quota license holder	
	% hunting	% successful	% hunting	% successful
Mail survey				
Residents	84	18	95	45
Nonresidents	93	36	95	47
Telephone survey				
Residents	80	13	89	40
Nonresidents	88	25	89	29

harvests. The check station is usually opened in late September or early October and typically remains open 24 hours a day until mid-November. All hunters passing on Highway 287 are required to report at the check station, whether en route to hunting, or returning from hunting, successful or unsuccessful.

Hunters must report the location of their kill by (1) specific drainage or geographical feature, (2) special management area, and (3) hunt area number. Also recorded are the date of the kill and its age and sex (mature bull, spike bull or yearling bull, cow, male or female calf).

Only elk transported east on US Highway 287 are subject to be monitored at the Dubois check station. The check station cannot monitor elk transported south toward Rock Springs or Star Valley, Wyoming, west to Idaho, north through Yellowstone National Park, or those flown out of the Jackson Hole Airport or killed by residents of Jackson Hole. In recent years, the human population in the Rock Springs area has increased substantially. There was concern that more elk killed in the Jackson herd unit were being transported south and that the Dubois check station was accounting for a smaller proportion of the total kill. To correct for this, Wood and Yorgason (1974) proposed a ratio estimator for the total number of elk killed in the Jackson herd which was based on a known kill of elk from Grand Teton National Park and the National Elk Refuge. Their estimator is

$$\hat{H} = h \ P/p \tag{3.5}$$

where $\hat{H}$ is estimated total harvest from the Jackson herd unit, h is the total number of elk checked at Dubois from the Jackson herd unit, p is the number of elk checked at the Dubois check station that were killed in Grand Teton National Park or on the National Elk Refuge, and P is the known kill of elk from the Park and Refuge.

This estimator assumes that the same proportion of elk killed in the Park and Refuge will be checked through the Dubois check station as are checked for the entire Jackson herd unit. This assumption may be valid in some years, but in most it is not. Total harvest from the Jackson herd unit estimated by mail survey and the harvest estimated by the model at equation 3.5 correspond

Table 3.8. *Nonresponse bias amongst Wyoming elk hunters for 1970.*

	Mature bulls	Cows	Spikes	Calves
Mail survey estimated kill	3441	6579	3281	1149
Telephone call estimated kill	3367	5946	2917	1003
Error	− 2%	− 10%	− 11%	− 13%

closely in a few years, but in other years there is substantial discrepancy (Figure 3.25).

Discrepancies between the two estimates are usually due to the timing of check station operation relative to the hunting seasons. In particular, if the Park and Refuge hunts continue past the date that the check station is closed, equation 3.5 will consistently overestimate the total harvest. If, however, the check station is open throughout the Park and Refuge seasons, but elk from other areas in the herd unit are missed either before or after the check station operates, the estimator at equation 3.5 consistently underestimates the harvest.

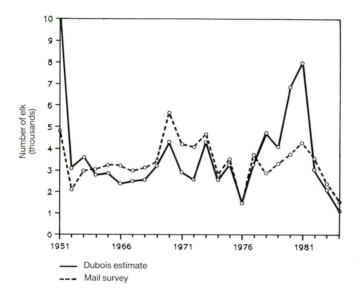

Figure 3.25 Total kill of elk from the Jackson herd unit estimated by mail survey and by the Dubois check station ratio estimator, 1951–52, 1963–84.

This temporal effect on the ratio harvest estimator is well illustrated by estimates from two recent seasons. In 1984, hunting seasons opened on 10 September in the Teton Wilderness and other parts of the Bridger-Teton National Forest. The temporal distribution of elk checked is illustrated in Figure 3.26 for (1) Park and Refuge elk, (2) Teton Wilderness elk, and (3) elk killed on other Forest areas. The Dubois check station was manned from 29 September through 21 November, although complete data are available only through 15 November. Any elk shot before 29 September or later than 21 November was not checked. In 1984, the Park and Refuge seasons opened 27 October and closed 12 November; therefore the check station was open throughout the period that Park and Refuge elk were likely to come through the check station. The seasons on all forest areas outside of the Teton Wilderness were closed by 2 November; therefore no elk were checked from these areas after the check station closed. But unquestionably elk were shot before 29 September, none of which would have been Park or Refuge elk. These unaccounted-for elk should have appeared in the numerator of equation 3.5, consequently the estimate of harvest is biased low.

The opposite situation occurred in 1981 when Park and Refuge hunts extended from 31 October through 6 December, but the Dubois check station closed on 16 November 1981. A substantial fraction of the kill from the Park and Refuge was removed after the station closed. Consequently, the denominator in equation 3.5 is smaller than it should be and harvest is greatly overestimated.

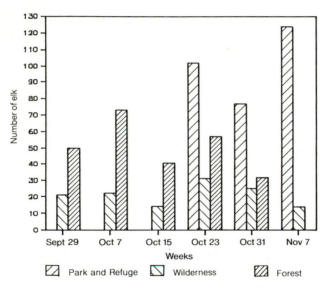

Figure 3.26 Harvest of elk from the Teton Wilderness, Park and Refuge, and other Forest areas by week for elk reported at the Dubois check station during 1984.

Other factors can bias the Dubois check station ratio estimator, although I suspect that the timing problem is most serious. One possibility is that Jackson residents may be more likely to participate in late migratory harvest simply because local hunters are better able to learn the timing of migration. This will bias check station estimates either up or down depending upon whether migratory elk were being shot in the Park or on the Forest.

To make valid use of harvest estimates derived from the Dubois check station records, the check station should be open during the entire period that any elk hunting seasons are open in the Jackson herd unit. Most years this would require that the station be open from 10 September through November, and sometimes into December. Maintaining personnel at the check station is expensive, however, and may not be justifiable. Closing the station at randomly selected dates throughout the season rather than at either end would reduce sampling bias. Even though the current system may yield the greatest number of animals checked, because of sampling bias, the larger sample may be less valuable than a smaller unbiased sample.

Harvests from the Park and Refuge

Hunters licensed to hunt on the National Elk Refuge or in Grand Teton National Park agree to report their kill to the respective agency. In the Park, check stations are established at Moose and Pacific Creek and hunters are requested to submit mandibles from their elk. For the most part, hunters appear to cooperate in reporting their kill from these special permit areas. However, there is always a fraction that do not. For example, on the National Elk Refuge, the fraction of hunters that failed to return their permits ranges from 7.6% in 1979 to 26.2% in 1977. Presumably these hunters were unsuccessful, but this is not certain, and some harvest may not have been tallied. For this reason, estimated harvest based on the mail survey sometimes do not coincide with the 'known' kill from the Park and the Refuge. Differences are usually minor, however, and are not considered to be large enough to merit closer study.

Crippling losses

Invariably, some elk are shot by hunters but not retrieved for various reasons. These crippling losses are usually presumed to be approximately 10% of the total harvest, but may be higher. In 1967, a questionnaire survey of guides and outfitters indicated a crippling loss of 22% (Straley, 1968). Also, guides and outfitters reported 44 elk found dead for every 100 that were recovered. During 1966, 3% of the number of elk on winter feedgrounds showed signs of being crippled, and most of these had obvious gunshot wounds (Straley, 1968).

Patterns in hunter harvest

Kill of cows and calves and the total kill of elk from the Jackson herd are highly correlated ($r = 0.93$, $n = 33$, $P < 0.001$) with a slope greater than unity ($b = 1.43$). This slope indicates that as the total kill increases in Jackson Hole, the proportion of cows and calves in the harvest becomes greater. Since 1946, the highest kill occurred in 1970, and the lowest during the no-snow year of 1976 when migration was delayed and conditions for hunting were poor.

There has been considerable concern about apparent recent declines in the

Jackson elk herd as evidenced by declining counts of elk on the National Elk Refuge (Figure 3.1) and on the Gros Ventre feedgrounds (Figure 3.2), although when standardized to a tough winter, the decline is not so apparent (Figure 3.3). It has been argued that overharvest in recent years has contributed to the decline (Melnykovych, 1985; Smith, 1985). However, comparing trends in total harvest, or in harvests of antlerless elk, it is not apparent that recent kills from the Jackson elk herd have been so large relative to earlier years. To highlight this, in Figure 3.27 I have plotted the five-year totals of elk harvest since 1950. The period 1970 to 1974 stands out as having the largest total kill, but total kills over the past 10 years do not appear out of line with average kills since 1950.

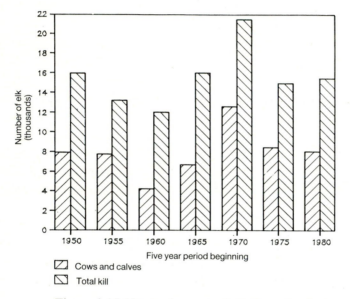

Figure 3.27 Hunter harvest of elk from the Jackson herd unit pooled into five-year intervals, 1950–84.

Spatial distribution of harvest has varied substantially from year to year. Harvest from the Park and Refuge increased substantially beginning in 1974 or 1975 (Figure 3.28) creating a significant trend in kills over the period 1960 to 1984 ($r_s = 0.677$, $n = 24$, $P < 0.001$). In 1981 there was a large kill of elk on National Forest lands, but this was not nearly as large as the sustained heavy kill of Forest elk from 1970 through 1974. A large fraction of the kill on Bridger-Teton National Forest during the early 1970s came from the Spread Creek area, but harvests from this area apparently decreased in recent years, although the overall trend from 1961–84 is not statistically significant ($r_s = -0.271$, $n = 23$, $P > 0.1$).

Kill of elk from Gros Ventre areas has shown substantial variation, but has

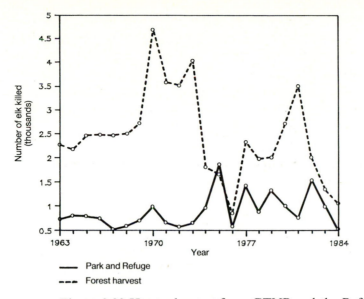

Figure 3.28 Hunter harvest from GTNP and the Refuge compared with that from all National Forest areas, 1963–84.

remained high. It is interesting to note that an enormous peak in the harvest from special management area 10 (Hunt Area 80; see Figures 3.22 and 3.23) occurred in 1970 (Figure 3.29), dominating kill statistics for the Gros Ventre that year. This harvest was poorly distributed, however. Most animals were shot along the western edge of the area, after coming off the National Elk Refuge where an early season had dispersed the elk (F. Petera, pers. comm.).

Elk shot in the Teton Wilderness are often elk migrating from Yellowstone National Park, particularly late in the season. When migration occurs during open hunting season, the harvest can be substantial, for example, in 1974 (Figure 3.30). Kill of elk from Teton Wilderness areas has declined significantly since 1963 ($r_s = -0.505$, $n = 23$, $P < 0.02$).

Consequences of harvest

Hunter harvest is a key factor determining the number, age and sex composition, and spatial distribution of the Jackson elk herd. In Chapter 2 we saw how migratory elk that formerly used the Togwotee migration corridor were drastically reduced in number, apparently in direct response to improved hunter access into the area and heavy harvests. Substantial variation in the sex and age composition of the population is in direct response to regulations controlling the composition of the harvest.

The most pronounced example of the impact of hunter kill on age and sex composition is the increase in the proportion of bull elk in Grand Teton

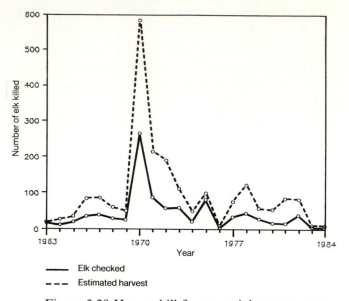

Figure 3.29 Hunter kill from special management area number 10 (see Figure 3.23). Kill from this area in 1970 is largely responsible for the enormous peak in harvests observed on the Gros Ventre for that year.

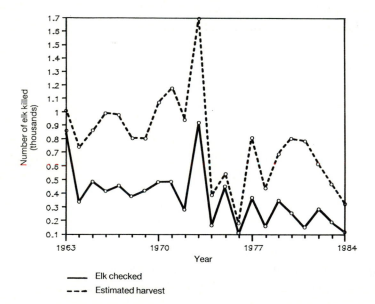

Figure 3.30 Hunter kill from the Teton Wilderness areas estimated from Dubois check station records, 1963–84.

National Park since 1976 as a consequence of regulations which resulted in Park harvests composed mostly of antlerless elk (Figure 3.31). This has had a dramatic effect on the composition of elk counted during the summer valley counts in the Park (Figure 3.20) and on the National Elk Refuge (Figure 3.19).

Harvest of elk from Grand Teton National Park is inversely correlated with Valley counts the following summer ($r = -0.435$, $n = 22$, $P < 0.05$). A very similar relationship exists between the kill rate of elk from the Refuge and the Park, and corrected counts of elk on the Refuge the following winter ($r = -0.442$, $n = 21$, $P < 0.05$).

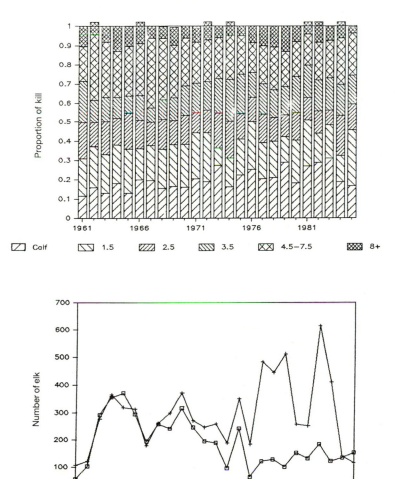

Figure 3.31 Age and sex composition of elk killed in Grand Teton National Park, 1961–85.

Below, I show that most of the year-to-year variation in the number of elk in the Jackson herd unit can be accounted for by recruitment and hunter harvest. This is not terribly surprising since most of the normal overwinter mortality is eliminated by the feeding program.

> Above we found that 50% of the variance in winter counts of elk on the National Elk Refuge is attributable to variation in January precipitation at Moose, Wyoming. Correcting these counts for January precipitation (equation 3.2), 76% of variation in the per capita growth rate (r_{NER}) from 1950 to 1985 is attributable to variation in hunter harvest and recruitment:
>
> $$r_{NER} = \ln(N_t/N_{t-1})$$
> $$= 1.636 \text{ (proportion calves}_{NER}) - 0.0004 \text{ (Kill}_{t-1} + \text{Kill}_t)$$
> $$+ 0.0001(N_t) - 0.002 \text{ (PCP}_{Jan}) - 1.195 \tag{3.6}$$
>
> All variables contribute significantly to this linear model with an overall F-ratio $= 11.375$ (5,18 df; $P < 0.001$). Two measures of recruitment contribute to population increase, both calves/100 cows in the preceeding harvest and the proportion of calves in Refuge winter counts. Total harvests of elk from the Jackson herd unit during the previous two hunting seasons is depicted $\text{Kill}_{t-1} + \text{Kill}_t$. Only about 12% of the variation in counts of elk on the Refuge remains to be accounted for, and may be mostly stochastic.
>
> A simpler model does nearly as well in accounting for variation in the pooled counts from 1960 to 1985 on the Refuge and the three State feedgrounds in the Gros Ventre valley. Here, 80% of the variation in per capita growth rate (r_{total}) is attributable to three variables:
>
> $$r_{total} = 0.021(\text{calves/100 cows}_{NER}) - 0.0004 \text{ (Kill}_{t-1} + \text{Kill}_t)$$
> $$+ 0.0003 \text{ (Bulls Killed)} - 0.475 \tag{3.7}$$
>
> Again, all variables contribute significantly to the model ($P < 0.001$) with overall F-ratio $= 20.8$ (3,16 df; $P < 0.001$). Both recruitment and harvest are again important variables, and here, the number of bulls in the harvest contributes significantly to variation in total counts. This reflects the effectiveness of bulls-only regulations for increasing herd growth rate (see Chapter 7).

Recent declines in the number of elk in the Jackson elk herd can largely be attributed to patterns of hunter kill. During the early 1970s, heavy kills depressed herd segments on National Forest lands (see Figures 3.28 and 3.30). Following this, heavy antlerless harvests from Grand Teton National Park and the National Elk Refuge successfully culled Grand Teton National Park elk and possibly those from western portions of National Forest and Yellowstone National Park lands (see Figures 3.6 and 3.31), while substantial kills were sustained on eastern Forest segments. Although the total kill did not increase between 1970 through 1982, (Figure 3.27), the cumulative effect of harvests over this period can account for the decreased herd size in recent years (Melnykovych, 1985).

Population regulation

Single-species models for sustained yield optimization require density dependent survival, recruitment or dispersal. Otherwise, populations should be allowed to build to enormous numbers to secure greater yields (Mendelssohn, 1976). Understanding of density dependence is important in establishing sustained yield harvest policies, and I think it would also help to anticipate probable consequences of the proposed experimental reduction or termination of the Park hunt (Grand Teton National Park, 1985).

Gross (1969) claimed that recruitment in elk populations is density dependent. I suspect that any such pattern largely reflects variation in neonate survival, since calf/cow ratios are often substantially lower than pregnancy rates (Greer, 1966). To evaluate density dependent recruitment I tested the null hypothesis that calf/cow ratios on the National Elk Refuge are not correlated with the previous winter's counts on the Refuge standardized for a tough winter. However, I am unable to reject the null hypothesis, and contrary to expectation, the correlation is weakly positive ($r = 0.279$, $n = 24$, $P > 0.1$). Supplemental feeding on the Refuge may ameliorate density dependent neonatal mortality, since density dependence in calf recruitment was clearly documented for the northern Yellowstone herd (Houston, 1982: 46).

Increased recruitment for elk in Grand Teton National Park since 1976 (Figure 3.9) may reflect reduced density as a consequence of increased Park harvest in recent years (Figure 3.28). No correlation occurs between calf/cow ratios in the Park and estimates of the total central valley population ($r = -0.021$, $n = 15$, $P > 0.5$); however, an inverse correlation exists between calf/cow ratios and estimates of the cow-calf-yearling male segment of the population ($r = -0.672$, $n = 13$, $P = 0.012$; Figure 3.32). This suggests that habitat segregation may exist for bull and nonbull segments of the population as implied by Martinka (1969) and demonstrated for red deer by Clutton-Brock, Guinness & Albon (1982). Also, a correlation exists between hunter harvest in the Park and recruitment during the following year (see below).

Picton (1984) found a density dependent relationship in the effect of climatic variation on elk recruitment. Climatic variation over the previous 18 months was correlated with calf/cow ratios in the upper Sun River drainage of central Montana, but only for five of 20 years when the elk population was above its estimated carrying capacity. I am unable to make such an assessment for density relative to the fluctuating carrying capacities for elk in Jackson Hole; rather, I simply selected those years for which Refuge counts corrected for winter severity were larger than one standard deviation above the mean (i.e., 1941, 1951, 1962, 1974, 1976, and 1978 where $N > 9230$). Similar to Picton's findings, calf/cow ratios on the Refuge the following season are positively correlated with annual precipitation at Moran ($r = 0.831$, $n = 6$, $P < 0.05$). Absence of this pattern when all years are included in the analysis ($r = 0.077$, $n = 29$, $P = 0.684$) suggests that the role of climate for elk recruitment is density dependent.

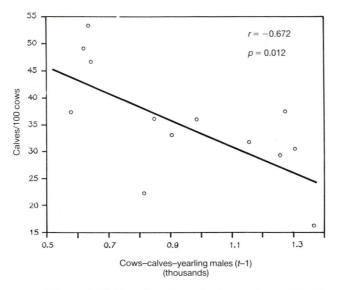

Figure 3.32 Density dependent recruitment in Grand Teton National Park, 1970–82.

For those years at large population size listed above, there exists a highly significant correlation between calf/cow ratios and per capita population growth rates $[\ln(N_t/N_{t-1})]$ for the Refuge herd corrected for winter severity ($r = 0.99$, $n = 5$, $P < 0.01$). However, this pattern is even more general: including all years for which data exist, we find a significant correlation between calf/cow ratios and the per capita growth rate in total feedground counts for the entire Jackson herd unit ($r = 0.638$, $n = 21$, $P < 0.01$).

To further evaluate the role of recruitment in regulating the elk population, I analyzed variation in recruitment rates (reflected by calf/cow ratios in the harvest) among hunting areas. Present hunt areas have only been in existence since 1971; therefore age-specific details of harvest span only the past 14 years. No estimates of population size for hunt areas is available; therefore I relied upon prior year's harvests to reflect reduced local population size from year to year. Since different hunt areas vary in size and in average harvest, I standardized all harvest statistics for each hunt area to mean $= 0$ and standard deviation $= 1$. Likewise, because recruitment rates may vary among hunt areas I have standardized these values for each hunt area as well.

I then fitted the following model

$$C'_{t,i} = \beta_0 + \beta_1 X'_{t-1,i} + \epsilon \tag{3.8}$$

where $C'_{t,i}$ is the number of calves per 100 cows in year t standardized $(0,1)$ for each hunt area i. Similarly, $X'_{t-1,i}$ is the number of elk killed in the previous year, $t-1$, standardized for each hunt area i. The standardization of variables allows hunt areas to be pooled to assess the consequence of previous harvests to relative recruitment rates over all areas and years. Because harvest regulations may preclude harvest of cows and calves, I eliminated all years in which regulations allowed only bulls to be shot and all years in which no calves were harvested.

Pooling all hunt areas over the past 14 years, we may reject the null hypothesis that relative magnitude of hunter kill is unrelated to recruitment the following year ($r = 0.222, n = 89, P < 0.05$). The relationship is even stronger for pooled hunt areas within the Park, the Refuge and the Teton Wilderness ($r = 0.396, n = 40, P < 0.01$). However, we cannot reject the null hypothesis for Spread Creek–Gros Ventre hunt areas ($r = 0.083, n = 49, P > 0.1$).

These results support Gross' (1969) premise that elk recruitment is density dependent. This analysis cannot assess, however, whether density dependent fecundity or survival of young calves is the primary component determining density dependence in recruitment.

Density dependence in the survival of female calves to yearlings also exists for elk in the Jackson area (Sauer & Boyce, 1983). Independent verification of density dependent calf survival is provided by the fact that male calf survival (Figure 3.13) is inversely correlated with the number of elk on feed at the National Elk Refuge ($r = -0.521, n = 21, P < 0.05$) as well as with the Refuge counts adjusted for a tough winter ($r = 0.453, n = 21, P < 0.05$).

But does this density dependent survival reflect natural regulation or might it simply be a consequence of the feeding program? Per capita quantity of feed provided at the National Elk Refuge is inversely correlated with the number of elk being fed at the Refuge ($r = -0.534, n = 24, P < 0.01$; see Chapter 5). However, this does not account for the density dependence in calf survival. Multiple regression of male calf survival as a function of both the number of elk on feed and the total per capita hay rations provided at the National Elk Refuge revealed that hay ration did not contribute significantly to the model ($t = 1.64$,

$P=0.117$), whereas the number of elk on feed maintains a strong influence on calf survival, even when hay ration is included in the model ($t=3.23$, $P=0.005$). This is despite the fact that per capita hay rations are confounded with counts on the Refuge (see discussion in Chapter 5). Furthermore, very few yearlings die on the Refuge during winter (Appendix G).

We conclude that both recruitment and juvenile survival are density dependent in the Jackson elk herd, despite the intensive management program. What of dispersal? For many species, dispersal can be an important population regulating mechanism, and among polygynous mammals, it is particularly pronounced among males (Dobson, 1982). Ramifications are important because one might postulate that termination of the hunt in Grand Teton National Park may not have particularly serious consequences if density dependent dispersal of yearlings allowed the Park population to naturally regulate despite the winter feeding program.

Evidence for regulation by differential dispersal of yearlings is implied by high frequencies of yearlings during summer on the National Elk Refuge and in valley areas of Grand Teton National Park, at the end of the period of apparent rapid expansion of the central valley herd segment (Martinka, 1969). The number of elk summering on the National Elk Refuge is small, and mostly restricted to wooded sections near the north end. In the summer of 1978, 125 of the 135 elk summering on the Refuge were yearling (spike) bulls (Unpublished National Elk Refuge Narrative Report, 1978). Spike/cow ratios from valley counts are positively, but not significantly, correlated with the number of elk summering on the Refuge ($r=0.444$, $n=13$, $0.05<P<0.1$).

The number of elk summering on the Refuge is inversely correlated with the corrected number of elk using the Refuge during the previous winter ($r=-0.532$, $n=22$, $P<0.05$). The fact that calf/cow ratios in Grand Teton National Park are positively correlated with the number of elk summering on the Refuge one year later ($r=0.696$, $n=13$, $P<0.05$) indicates that high recruitment years with low density are likely to be followed by years in which there are more yearling elk summering on the National Elk Refuge.

Martinka (1969) suggests that the number of yearling males and females remaining on the Refuge and in the valley areas fluctuates from year to year, because in some years a substantial fraction of the yearlings do not migrate. He observed one marked male in mountain areas with migratory elk as a calf and then later in valley areas of the Park as a yearling. Spike/cow ratios during summer were highest in southern areas of the valley, and tended to decrease further north (Martinka, 1969: 470). We do not adequately understand the mechanisms influencing yearling dispersal in elk, but Martinka's discussions and the density dependence in Refuge summer counts imply that when conditions are good, yearlings may remain in low elevation areas near wintering ranges rather than migrating. This suggestion is reinforced by a positive correlation between June precipitation in the valley and the number of elk summering on the Refuge ($r = 0.55$, $n = 22$, $P < 0.01$).

An alternative hypothesis to explain the high proportion of yearlings in the central valley is that orphaned calves are less likely to migrate. If this were true, we expect the proportion of spike bulls in valley counts to be positively correlated with the proportion of cows culled in the previous autumn. However, I am unable to reject the null hypothesis of no correlation between the proportion of cows in the total herd unit harvest and spike/cow ratios in the central valley areas of the Park during the following summer ($r = 0.012$, $n = 17$, $P > 0.5$). Likewise, the correlation between the proportion of cows in the Park harvest and subsequent spike/cow ratios in the Park is not significant ($r = 0.285$, $n = 12$, $P > 0.1$).

To assess whether dispersal of spikes might be density dependent and influenced by previous hunter kill, I repeated the standardized-variable regression (equation 3.8), but substituted spike/cow ratios for calf/cow ratios. I found no evidence whatsoever for density dependence in spike/cow ratios among harvested elk. Likewise, in an unhunted population on the Isle of Rhum, Clutton-Brock, Major & Guinness (1985) found no evidence for dispersal regulation of red deer into an adjacent area where culls were conducted. Still, differential dispersal of yearling males occurs in the valley, and Long *et al.* (1980) documented a young radio-collared cow elk that changed summer ranges.

It can be difficult to demonstrate density dependence unless population size has approached carrying capacity (Slade, 1977). Populations fluctuating at densities below carrying capacity may not have experienced the limiting forces that would regulate a population at higher densities (Fowler, 1981). We have demonstrated density dependent recruitment and survival that is independent of feedlot effects, although it is not known when during the year that these density dependent effects occur.

Yield optimization

Per capita growth rate is usually a concave function of population size among vertebrates (Fowler, 1981, 1987; Kie & White, 1985). This implies a population growth model which is structurally more complex than the logistic model, with the simplest possibility being the Gilpin & Ayala (1973) model:

$$dN/dt = r_0 N(1 - [N/K]^\theta) \qquad (3.9)$$

where r_0 is the potential growth rate with no density dependent effects (i.e., when $N = 0$), K is carrying capacity, and θ determines the relative concavity of dN/Ndt as a function of N.

For $\theta > 1$, the yield curve obtains a global maximum at $N > K/2$. Data on

harvests and population size allow estimation of yield curves. We approximate
dN/dt at equation 3.9 to be approximately the change in population size plus
hunter kill. For elk summering in central valley portions of Grand Teton
National Park (Table 3.2) being harvested in the Park and on the National Elk
Refuge, the evidence for a yield curve is convincing (Figure 3.33). We approximate the yield curve parameters by least squares to find a carrying capacity, i.e.
where $dN/dt = 0$, equal to 2043 elk. For this model, the maximum sustained
yield (MSY) is 910 elk per year at a valley population size of 1329. This high
yield is obtainable because the total population being harvested in the Park and
the Refuge is actually much larger than just that in the central valley area of the
Park.

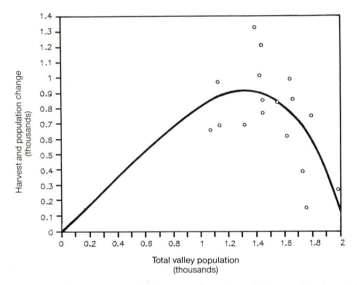

Figure 3.33 Yield curve for Grand Teton National Park, based upon
hunter harvests from the Park and the Refuge and estimates of the
central valley summer population of elk in the Park. Parameters for the
Gilpin–Ayala model (equation 3.9) are $K = 2043$, $r_0 = 0.881$, and
$\theta = 3.5$.

Although there is perhaps inadequate data on this population whilst near its
carrying capacity, the model illustrated in Figure 3.33 suggests that with no
hunting the population will eventually stabilize with approximately 2000 elk in
the central valley population. Harvest levels during the past 25 years have
maintained the population slightly above its maximum sustained yield density,
with valley estimates averaging 1510 elk.

A yield model is also possible based upon winter counts of elk at the National
Elk Refuge corrected for a severe winter as at equation 3.2, and harvests of elk

from the Jackson herd since 1950 based upon mail survey estimates. Here again, the fit to a Gilpin–Ayala model is highly significant ($P < 0.001$). Least squares nonlinear regression estimates for $r_0 = 0.636$, $K = 11\,225$, and $\theta = 3.027$. These parameters suggest a maximum sustained yield of 3387 elk at a Refuge population of 7100 elk (Figure 3.34). As was true for the Park segment, the Refuge population has averaged higher than maximum sustained yield at 8325 elk when corrected for a severe winter, or 7110 elk actually counted on feed. Actual yields have averaged 2913 elk since 1950, which is 86% of maximum. Again, a yield of nearly 48% of the Refuge population is possible because the total kill encompasses elk using the three Gros Ventre feedgrounds or wintering off of any feedground in the valley. Therefore, the yield value comes from a larger population than uses the National Elk Refuge.

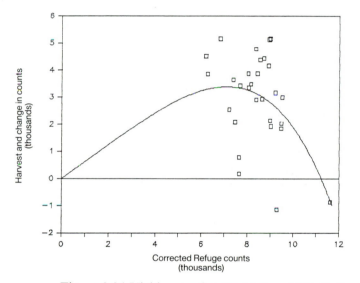

Figure 3.34 Yield curve for the National Elk Refuge based upon harvests from the entire Jackson herd unit and counts from the National Elk Refuge, corrected for a tough winter. Parameters for equation 3.9 were estimated by least squares to be $r_0 = 0.636$, $K = 11\,225$, and $\theta = 3.027$.

Behavior of the Refuge population near carrying capacity is even more tenuous than for the Park population. If we assume that the Gilpin–Ayala model provides a reasonable characterization of the yield curve, the projected carrying capacity is 11 225 given the current feeding program and density dependence in survival and recruitment.

Pooling corrected counts from the Refuge with counts of elk from the 3 Gros Ventre feedgrounds, we can attempt to find a yield curve for the entire Jackson elk herd (Figure 3.35). Unfortunately, the plot is not very convincing. There is so much variance that a quadratic does as well as a cubic polynomial in characterizing yield versus total winter counts. Likewise, $\theta = 1.0$ does as well as $\theta > 1.0$ at reducing the residual variance in estimates of the Gilpin–Ayala model. Being unable to correct Gros Ventre counts for winter severity appears to enter excessive variance into these pooled counts.

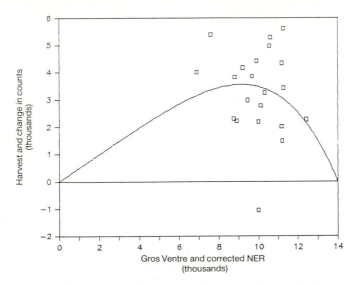

Figure 3.35 Yield curve for the Jackson elk herd, based upon the sum of corrected Refuge counts and total counts on the three Gros Ventre feedgrounds. Harvests are totals for entire herd unit estimated from mail surveys. By forcing $K = 14000$, least squares estimates for $r_0 = 0.498$ and $\theta = 3.557$ as at equation 3.9.

Arbitrarily forcing $K = 14000$, I estimated $r_0 = 0.498$ and $\theta = 3.557$ by least-squares nonlinear regression of equation 3.9. Maximum sustained yield of 3551 elk obtains at a population of 9950 elk. Maximum sustained yield for the Refuge segment of the population is similar to that for the combined Refuge and Gros Ventre feedground population. For a post-season herd of 11 000 at herd unit objectives, this implies a maximum sustained yield of nearly 30% of the population. I suggest that this is higher than actual because there are more elk wintering off feed than estimated.

Summary

1. Greater numbers of elk attend winter feedgrounds during years of heavy winter precipitation.

2. The number of elk attending the National Elk Refuge during winter has increased slightly since feeding commenced in 1912.

3. Overall, the sex ratio of calves in hunter harvests is 1:1; however, in recent years, a preponderance of female calves have been killed in Grand Teton National Park during autumn. This is correlated with an increase in the proportion of mature bulls in the Park consequent to antlerless elk culls.

4. Recruitment of calves averages greater than 30 calves per 100 cows in mid-winter, which is similar to that for other elk herds in western North America.

5. Nearly 90% of the year to year variation in the number of elk at the National Elk Refuge during winter can be attributed to winter severity, recruitment and hunter harvest.

6. Recruitment and survival decline as population density increases despite the winter feeding program. Also, yearlings are less likely to migrate to high-elevation summer ranges when densities are low and summers are wet.

7. The maximum sustainable yield of elk for hunters is approximately 3500 elk.

Chapter 4.

Habitat ecology

Successful management of elk requires knowledge of habitat use as well as the effects of elk on vegetation. Thus, it is appropriate that habitat ecology was a chief concern in past studies of the Jackson elk herd (Anderson, 1958; Cole, 1969). Continued vigor of the Jackson elk herd depends upon maintaining suitable habitats. Impending development, continued livestock grazing pressures, logging, road construction, and perhaps even the elk themselves potentially threaten habitats, therefore careful planning and management of habitats is critical.

Most elk in the Jackson herd are migratory with summer-fall ranges very different from those used during winter and spring. In this chapter I begin with a review of important seasonal habitats, including discussions of migration habitats and spring-time calving areas. In the last sections of this chapter I develop in more detail two key habitat issues of significance in management of the Jackson herd: (1) habitat interactions with other ungulates, especially cattle, and (2) the complex relationships between elk and aspen.

Summer-fall ranges

Summer range for elk is often thought not to be limiting to elk because it is winter range that usually constrains elk population size (Lyon & Ward, 1982: 470), particularly in regions of heavy snow accumulation. Summer ranges for the Jackson elk herd are extensive and appear to harbor virtually unlimited supplies of forage. Still, it has been argued that summer habitats are important in determining the population biology of elk (Casebeer, 1961), and that elk may seriously degrade these habitats (Beetle, 1979). Such degradation may include reducing aspen and other browse plants, increasing the frequency of annual grasses and weeds, and increasing soil erosion.

Generally, the role of summer range in the population biology of wild ungulates has been inadequately studied; however, there is evidence that summer nutrition can have important consequences to reproductive performance, growth and subsequent survival. Klein's (1965) studies of black-tailed deer in Alaska indicate that summer forage quality influences reproductive success, growth and overwinter survival. Summer nutrition has been shown to influence reproductive performance among moose (*Alces alces*; Edwards & Ritcey, 1958), mule deer (*Odocoileus hemionus*; Julander, Robinette & Jones, 1961; Hungerford, 1970), and white-tailed deer (*O. virginianus*; Verme, 1963, 1965, 1967; Ransom, 1967). Nutrition on summer ranges can influence calf growth rates and subsequent overwinter survival in reindeer (*Rangifer tarandus*; Reimers, 1972). Likewise, Geist (1971) found that summer range affected maintenance of reproductive condition and survival during the following winter in bighorn sheep (*Ovis canadensis*).

Some authors have contended that summer ranges within the Jackson herd unit have been 'overgrazed' by elk (Croft & Ellison, 1960; Beetle, 1974a, b; Casebeer, 1961), both on the Yellowstone–Teton Wilderness high-elevation ranges and on the valley ranges in Grand Teton National Park and along the Gros Ventre River (Beetle, 1974a). Others have insisted that there is no evidence that most of these ranges are overgrazed (Gruell & Loope, 1974), and these summer ranges will continue to sustain a sizable elk herd. In a range management context, 'overgrazing' means that continued grazing at a particular stocking rate will create changes in vegetation composition (Stoddart & Smith, 1943). The worst cases of overgrazing occur for introduced herbivores, for example, elk and sheep on New Zealand and rabbits and goats on Mauritius, which may ultimately result in loss of vegetation and escalated soil erosion (Pimm, 1987).

Mountain ridge habitats

During summer and fall, elk in the Jackson herd make extensive use of high-elevation herblands, particularly along several mountain ridges in the northern Teton Wilderness and southern Yellowstone National Park. Cole (1969) documented general patterns of elk use of different vegetation types on mountain ridges during the summers of 1962–66 (Figure 4.1). Such observational data are often limited to daylight hours in relatively open cover types. This creates a bias against elk use in heavy cover and during nighttime (Skovlin, 1982).

Mountain ridge habitats in the Jackson herd unit were investigated several times during the 1950s (Beetle, 1952; Croft & Ellison, 1960). Croft & Ellison (1960) report that the ranges appeared to have been severely depleted by overgrazing and trampling by elk, although they conclude that more investiga-

tion was required. A subsequent report by Beetle (1962) supported this interpretation. Since cattle have never grazed these ranges, Beetle concluded that elk
were responsible. The implication was that the National Elk Refuge feedground was sustaining a higher population of elk than the summer ranges could
support or than they supported prehistorically. Data presented in the previous
chapter suggest that this view was wrong since the range has certainly continued
to sustain the elk herd, indeed range condition may have improved in some
areas (see Chapter 6). It is not clear from any of these early range surveys what
grazing levels should be considered excessive. Curiously, Croft & Ellison
explicitly noted that reduction in the elk herd will probably have little consequence in alleviating range condition, because of poor site potential.

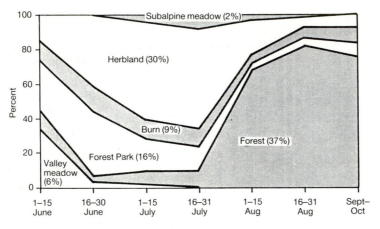

Figure 4.1 Use of vegetation types in high-elevation summer habitats
recorded by Cole (1969), based upon 20017 sightings of elk between
1962 and 1966. Sample sizes range from 641 for 1–15 June to 6543 for
16–31 July.

Casebeer (1961) concurred that summer range on Big Game Ridge was in
poor condition. A survey of 90% of the area used by the Jackson elk herd which
was completed in 1950 suggested that 27% was in fair condition, and 52% was
in poor or depleted condition. None of the area was classified to be in better
than fair condition, and accelerated soil loss was noted on 21% of the 92308 ha
surveyed.

Estimates of forage production and carrying capacity on Huckleberry Ridge,
in the northwestern portion of the herd unit, were made by Guest (1971).
He concluded that herbaceous forage production ranging in excess of 500 kg/ha
was certainly adequate to sustain the elk population in the area, although he
noted that elk concentrations early in the season caused localized soil erosion
and trailing.

An exclosure to exclude elk was constructed on Big Game Ridge in 1960 and an additional one on Gravel Creek in 1962. Vegetation was measured inside and outside of these exclosures when they were established and again in 1966, but no apparent trends were observed (Gruell, 1973) during this brief period. Gruell concluded that earlier investigators misinterpreted the causes of the conditions of these ranges because he failed to find evidence that overutilization and trampling by elk were the cause of presumed accelerated soil erosion.

Likewise, Gruell (1973) questioned whether range condition was deteriorated as claimed by Croft & Ellison (1960), Casebeer (1961), and Beetle (1974a, b). Gruell notes that indeed ground cover is sparse in some areas, but these are sites with poor soil due to the geological features and microclimates associated with the sites. Gruell argued that pocket gophers contributed to apparent poor range condition by increasing the occurrence of plants characteristic of disturbed sites (Laycock & Richardson, 1975). He presents numerous photographs spanning 99 years (1872–1971) which indicate that these ranges showed an increase in vegetation over that period.

Plant communities on Big Game Ridge are primarily herblands or tall forb communities. Forbs rather than grasses dominate; therefore criteria for judging the conditions of grassland ranges are difficult to apply to these communities. Gruell (1973) demonstrated that forage use by elk was positively correlated with the productivity of the site, and low productivity sites were little used.

Tracking (trailing) is heavy on many high-elevation elk ranges (Figure 4.2), and such impacts are most severe in areas where elk arrive in spring or early summer during snow melt (Guest, 1971). Gruell (1973) dismisses concern over tracking by suggesting that 'tracking is inherent and inevitable, [and] it should be accepted as a natural biotic effect'. He questions whether reductions in elk populations could improve range on Big Game Ridge given its 'poor hydrologic condition'. Gruell concludes by recommending that pellet group transects be measured each September and that quadrats and photo-point data be collected periodically to confirm results of his analysis and to document trends. Unfortunately, his work has not been continued.

Gruell's (1973) assessment of the consequences of range use by elk is similar to that of Houston (1982) for ranges in northern Yellowstone National Park. Basically, they agree that vegetation is altered by the presence of elk, but this is natural and should not cause concern. Semantic debates have developed over whether the range condition should be considered to be in a state of 'zootic climax' (Cayot, Prukop & Smith, 1979), 'biotic disclimax' (Cole, 1969), or 'zootic disclimax' (Beetle, 1974b).

I fail to find much of interest in this semantic debate. I agree with Cayot *et al.* (1979) that deterioration of range sites has not been convincingly demon-

Figure 4.2 Heavy trailing caused by horses and elk on Huckleberry Ridge in the Teton Wilderness. Photo by M. S. Boyce, July 1985.

strated. But at the same time, I disagree with Cayot *et al.* and Houston (1982) that there is any evidence for an equilibrium interaction between elk and vegetation (Caughley, 1976). There is inadequate evidence that such an equilibrium exists; rather the interaction between elk numbers and vegetation may yield perpetual fluctuations.

Such notions as 'dynamic equilibrium' have been suggested to account for the fact that fluctuations in population size appear to be the rule rather than the exception (Houston, 1982). However, a 'dynamic equilibrium' typically assumes a simple model which still possesses a stable equilibrium. The only reason that it is a 'dynamic' equilibrium is recognition that stochastic perturbations to the system can maintain fluctuations. Although this dynamic equilibrium model may be appropriate, a dynamic system can also exist which is truly dynamic, i.e., where there is no equilibrium. Sustained periodic oscillations are possible in seasonally forced plant–herbivore systems (Inoue & Kamifuku-moto, 1984; Schaffer, 1985) such as those which we have documented for elk wintering on the National Elk Refuge (Sauer & Boyce, 1979b).

Although the Forest Service has not continued range studies on mountain ridge summer ranges, the National Park Service has estimated forage utiliza-tion by the grazed/ungrazed plant method (Roach, 1950) most years between 1967 and 1983 on Red Creek Ridge in southern Yellowstone National Park, which is more accessible than Big Game Ridge. Proportion of plants utilized is estimated from a graph (Cole, 1963) or directly from the linear function:

$$\text{utilization} = (\text{proportion grazed}) \, (0.874) - 0.074 \qquad (4.1)$$

Measurements were taken at the end of the elk summer-range use period, as early as 29 August in 1975, and as late as 4 October in 1972 (R. P. Wood, unpublished).

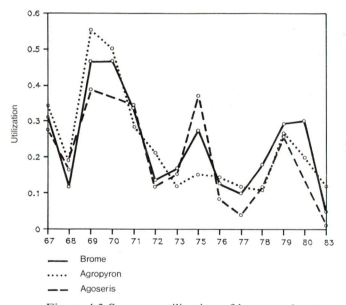

Figure 4.3 Summer utilization of brome, wheatgrass and false dandelion on Red Creek Ridge in southern Yellowstone National Park, 1967–83.

On Red Creek Ridge, primary forage plants are (1) mountain brome (*Bromus marginatus*), (2) slender wheatgrass (*Agropyron trachycaulum*), and (3) false dandelion (*Agoseris* sp., either *A. aurantiaca* or *A. glauca*; Shaw, 1976). Utilization values for each of the species are highly correlated among years (Figure 4.3), for example, between mountain brome and false dandelion, $r = 0.926$. Likewise, there are no significant differences in the distribution of utilization for any of the three plant species over years (Kolmogorov–Smirnov max. diff. <0.25, $n=14$, $P>0.214$). Levels of utilization on these species are virtually identical with measurements reported by Guest (1971) on Huckleberry Ridge in the Teton Wilderness.

There is, however, a declining trend in utilization of these plants by elk since 1967, which is statistically significant for slender wheatgrass ($r_s = -0.603$, $n=14$, $P<0.05$) and false dandelion ($r_s = -0.599$, $n=12$, $P<0.05$), but not quite so for mountain brome ($r_s = -0.418$, $n=14$, $P>0.1$). The patterns suggest little selection by elk for one species over another. Also, high consistency in utilization rates among plants within years, but substantial variance among years suggests that elk use of the area sampled by the transects varies substantially from year to year. The declining trend may imply a decrease in the number of elk using the area. Alternatively, if plant biomass fluctuates markedly from year to year, utilization rates may also be expected to be highly correlated irrespective of elk numbers.

Forest habitats

Well over one half of the Jackson elk herd spends the summer in forested areas, either in Grand Teton National Park or on Bridger–Teton National Forest. I will focus on National Forest lands here, and discuss habitats in the Park in the following section.

National Forest

National Forest lands are managed on a principle of 'multiple use', whereby the Forest Service attempts to balance numerous demands for public lands including livestock grazing, timber harvest, mining, recreation and hunting. Forest Service guidelines for management of elk attempt to identify elk habitat requirements, particularly when attempting to evaluate the consequence of timber harvest. In western North America the Forest Service approach to meeting elk habitat requirements is to manage for desirable mixtures of forest and open habitats, and to minimize road density (Lyon, 1979; Thomas, 1979; Leege, 1984).

Elk are presumed to be 'fine-grained', i.e., broad-ranging, in their choice of habitats, and detailed site-specific habitat information does not improve the ability to characterize use of habitats by elk. Rather, elk habitat evaluation can be on a larger scale allowing one to characterize summer habitats quite adequately from maps and aerial photographs. One of the most common approaches is to identify a boundary around roads (Lyon, 1979; Thomas, 1979), and to characterize gross habitats on the basis of forage/cover ratios where cover areas are defined to be forested areas and forage areas are nonforested areas (Leege, 1984).

On Bridger–Teton National Forest increased logging during the 1960s and 1970s prompted a joint Wyoming Game and Fish – US Forest Service research project known as the Gros Ventre Cooperative Elk Study, which was conducted from 1974 to 1979 (Long *et al.*, 1980). Research focused in the vicinity of the North Fork of Fish Creek on Bridger–Teton National Forest east of Jackson. The objective was to monitor the consequences of logging development in the area.

Twenty-one elk were monitored with radio transmitters during the study, and a total of 1700 relocations of these elk was recorded. As in other studies on the effects of logging (Hieb, 1976; Irwin & Peek, 1979, 1983b; Edge & Marcum, 1985; Edge, Marcum & Olson, 1985), it was found that elk are sensitive to disturbance and are displaced by logging activities. Removal of timber was not necessarily a deterent to elk use so long as disturbance could be minimized, and elk reoccupied logged areas after timber harvest activities ceased (Gruell and Roby, 1976).

Logging and associated road development has been identified in recent years to have extremely important consequences to elk habitat in the western United States (Hieb, 1976; Thomas, 1979). Open roads apparently have greater adverse impact than logging itself. Elk were displaced an average of 2.7 km from well-traveled roads in the Gros Ventre area, whereas infrequently traveled or closed roads resulted in a mean displacement of 1 km. Activity on the road and not the mere presence of the road displaced elk (Gruell & Roby, 1976). Average distance between elk and human activity averaged 3.2–4.2 km during 1976–79 in the Gros Ventre study. The key variable determining the response by elk to human activity appears to be sight distance, which can vary substantially due to the openness of the habitat (Thomas, 1979). Disturbance typically created a 0.5–1 km elk-free buffer zone around the site of human activity, which can substantially reduce available habitat for elk (Edge & Marcum, 1985).

Patches of timber on elk range should be no smaller than 10 ha to offer adequate escape cover, and somewhat larger patches up to 26 ha are ideal (DeByle, 1985b). The importance of thermal cover for elk (Thomas, 1979) has recently been challenged (Peek *et al.*, 1982; Peek & Scott, 1985), but no one questions the need for adequate escape or hiding cover. Typically conifers offer better escape cover for elk than aspen (Thomas, 1979). In particular, decadent stands of aspen (e.g., Figure 4.22) offer poor escape cover and virtually no thermal cover in winter (DeByle, 1985c), although they may provide shade.

During the Gros Ventre studies, elk used areas with 52% cover and 48%

forage, both in logged and unlogged areas. Because most elk use is within 183 m of edges between cover and forage areas (Reynolds, 1966; Harper, 1969), Long *et al.* (1980) adjusted their estimates of cover/forage ratio to exclude habitat patches further than 183 m from edge. This resulted in an estimated 54%/46% cover/forage ratio for areas used by radio-marked elk in the Gros Ventre.

These cover/forage 'ratios' are higher, i.e. with more cover, than for the Blue Mountains in Oregon, where optimal cover/forage ratios are thought to be approximately 40%/60%. In a study in forested habitats of Poland, Bobek, Boyce & Kosobucka (1984) found the 40%/60% ratio to be approximately optimal for red deer as well. One might speculate that the Oregon and Poland studies were in denser forest than the Gros Ventre area, thus the animals may have required less cover than elk in the Gros Ventre. However, because studies of elk cover/forage ratios have not been adequately replicated, these cover/forage ratios may not differ statistically – or even ecologically. The key point is that optimal elk habitat contains a balanced mixture of forest and open habitats with high interspersion of both and a high perimeter of forest edge.

Grand Teton National Park

Habitats for elk within Grand Teton National Park are mostly in the central valley portions of the Park, west of the Snake River (Figure 8.2). Very few elk wander into the Teton Range itself; rather, the greatest concentrations of summering elk are immediately in front of the Tetons in the vicinity of Timbered Island, the Snake River bottoms and Signal Mountain. Although very little of the park is classified as elk winter range (see Figure 2.3), portions of the Park are used by elk in late fall and early spring during most years. Also, during mild winters of little snow accumulation, portions of the Park may be used during most of the winter. Although much of the Park is forested, elk make extensive use of grasslands, sagebrush steppe and valley meadows.

During his tenure as a biologist for Grand Teton National Park in the 1960s, Glen Cole compiled data on general vegetation types for 82 223 elk observations (Figure 4.4). Most observations were recorded in open habitats; however, these data are biased against habitats where visibility is low, and against those used more extensively at night. Cole (1969) also reviewed elk food habits, but I am not aware of any subsequent or more detailed investigations on the food habits of elk in Jackson Hole.

As reported in Chapter 3, few elk were present in Grand Teton National Park during summers prior to 1958. By 1963, there were over 1500 elk in central valley portions of the Park, and by 1987, nearly one half of the Jackson herd summers in the Park. Speculation on mechanisms that led to the increase in elk in the Park has focused on (1) elimination of cattle grazing west of the Snake

River within the Park, and (2) a ban on hunting west of the Snake River within the Park.

Observations by Murie (1951c) and Guest (1971) that harassment by biting insects, for example, horseflies (Tabanidae), is lower on high-elevation ridges implies that insects may be partly responsible for seasonal migrations out of the valley. Indeed, insect harassment can be a serious annoyance to elk (Collins & Urness, 1982; Cole, 1969). However, if it was responsible for migration, it seems

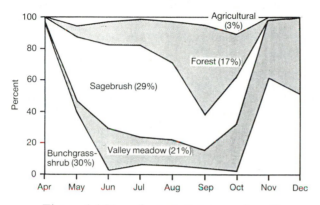

Figure 4.4 Use of vegetation types in valley areas of Grand Teton National Park recorded by Cole (1969), based on 82 223 sightings of elk during 1962–66. Monthly sample sizes range from 5525 to 19 398 elk.

contradictory that the elk population west of the Snake River in the valley increased substantially (Cole, 1969; Martinka, 1969) subsequent to removal of hunting in 1950, and cattle grazing by 1958. It is not possible to determine the relative importance of hunting versus cattle grazing in suppressing elk use of the Park. Certainly both factors can reduce elk use of an area. Yet, displacement of elk by cattle ranching seems to me the most likely explanation for increased summer use of the Park since hunting does not occur in the Park until autumn.

Between 1963 and 1974, transects were established east of the Snake River in Grand Teton National Park to assess grass and browse use by elk and other ungulates. Eight three-step transects were established to monitor range use, with focus on elk use of bluebunch wheatgrass (Figures 4.5, 4.6, 4.7). Methodology was identical to that employed on the Red Creek transects discussed above in conjunction with equation 4.1. Transects were run in May each year before new growth obfuscated the use of the previous year's vegetation. These transects are located in areas which do not sustain summer elk use that is as heavy as in areas west of the Snake River. Also most of these transects are subject to elk use in autumn and spring.

Percent utilization of bluebunch wheatgrass was lowest (mean = 19%,

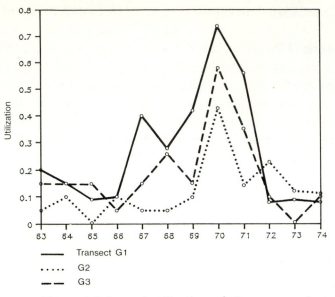

Figure 4.5 Annual utilization of *Agropyron spicatum* along three transects surveyed during May on Blacktail Butte, 1963–74.

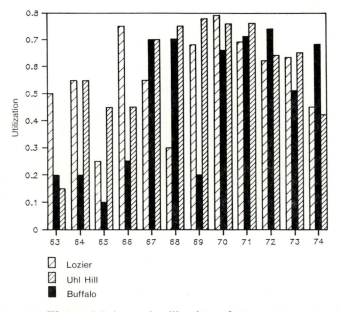

Figure 4.6 Annual utilization of *Agropyron spicatum* along transects surveyed during May on Lozier Hill, Buffalo Hills, and Uhl Hill, 1963–74.

range $= 0 - 75\%$) on Blacktail Butte (Figure 4.5), which is immediately north of the National Elk Refuge. Utilization increased significantly over the period 1963–74 on transect G2 ($r_s = 0.716$, $n = 12$, $P < 0.02$), but not on the other two transects on Blacktail Butte. Nevertheless, utilization on the three transects is highly correlated, suggesting some consistency in the measurement of utilization for the area. The reason for the pronounced peak in utilization on all three transects in 1970 may be the delayed migration of elk onto the National Elk Refuge during autumn 1969 (Table 2.2). More extensive use of areas adjacent to the Refuge occurred in part because 1969 was an exceptionally dry autumn.

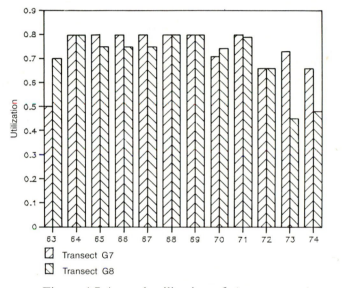

Transect G7

Transect G8

Figure 4.7 Annual utilization of *Agropyron spicatum* along 2 transects surveyed during May in the Spread Creek area of Grand Teton National Park, 1963–74.

Utilization of bluebunch wheatgrass on Lozier Hill, Buffalo Hills and Uhl Hill averaged approximately 50% during 1963–74. Utilization along the Buffalo Hills transect, which enjoyed low use during years 1963–66, increased significantly over the 1963–74 period ($r_s = 0.617$, $n = 12$, $P < 0.05$). Increasing use by elk may be due to decreasing cattle use during the sampled period. Particularly in spring, elk now use ranges further south than when heavier cattle use occurred in this area (R. P. Wood, unpublished). Distributions of utilization on these three transects did not differ significantly (Kolmogorov–Smirnov max. diff. < 0.333, $n = 12$, $P > 0.107$), and utilization levels between transects were not significantly correlated ($P > 0.1$).

Highest use of bluebunch wheatgrass is seen on the two Spread Creek transects (Figure 4.7), where mean annual utilization (May) is greater than

70%. The level of utilization on these two transects was consistently high during 1963–74 and was significantly higher than utilization on any of the other six range transects ($P < 0.02$). In this area, *Agropyron spicatum* is used mostly by elk during fall and spring.

Three transects (Uhl Hill, Spread Creek G7, and Spread Creek G8) have been measured in autumn as well as spring to obtain an index of summer use of bluebunch wheatgrass by cattle (R. P. Wood, pers. comm.), although some elk also use the areas in summer. Use during summer varies greatly, apparently due to variation in cattle grazing from year to year (Figure 4.8). High skewness in the distribution of utilization from Uhl Hill (skewness = 2.53) and Spread Creek G7 (skewness = 2.15) indicates high utilization some years, but low use during most. These transects are on slopes which are used typically only by cattle if lower areas are being overgrazed (R. P. Wood, pers. comm.).

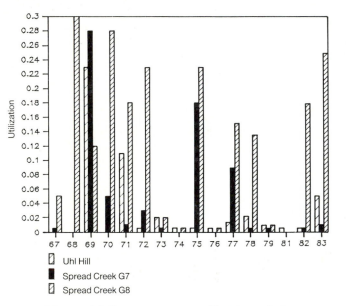

Figure 4.8 Summer range utilization of *Agropyron spicatum* along transects surveyed during autumn on Uhl Hill and two sites in the Spread Creek valley of Grand Teton National Park, 1967–83. Summer grazing pressure along these transects is mostly due to livestock.

Perhaps in part due to this high variance, there were no significant trends in utilization (mostly by elk) on these three transects during 1967–83. Even though the transects are near to each other, there is no correlation between utilization estimates on Spread Creek transects G7 and G8. However, utilization on transect G8 is significantly greater than on transect G7 ($P < 0.001$). Utilization at Uhl Hill is not significantly different from that at G7 ($P = 0.921$), but lower than at G8 ($P < 0.001$).

The three transects measured both in the spring and fall show substantial differences in utilization at these times. The majority of summer use on Uhl Hill and Spread Creek transects is from cattle, but it averages only 7%. In contrast, utilization on these transects by elk from fall through spring averages 68%, and is significantly higher than the use by cattle during summer ($P < 0.001$).

Browse can be important in elk diets, particularly when snow depth exceeds 50 cm (DeByle, 1985c) and covers herbaceous vegetation (Nelson & Leege, 1982) or in autumn when herbaceous vegetation dries out. Cole (1969) established several permanent plots to monitor the use of browse, condition and year-to-year trend of serviceberry (*Amelanchier alnifolia*), and bitterbrush (*Purshia tridentata*) in conjunction with the *Agropyron spicatum* transects described above. Although he notes that Douglas rabbitbrush (*Chrysothamnus viscidiflorus*) was also heavily used at times, he did not measure this species because it was very abundant and appeared able to withstand grazing. Methods of measuring condition and leader use are described in Cole (1963). Data were collected in May or June.

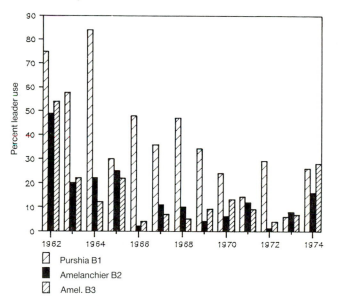

Figure 4.9 Browse leader use estimated in May for *Purshia tridentata* and two samples of *Amelanchier alnifolia* on Blacktail Butte in Grand Teton National Park, 1962–74.

Browsing ungulates forage mostly on the 'leaders' of shrubs. Leaders constitute the past year's growth, which is least woody and thereby most nutritious. Percent leader use for serviceberry and bitterbrush declined between 1962 and 1974 at Blacktail Butte (Figure 4.9), Lozier Hill, Uhl Hill (Figure 4.11), and Buffalo Hills (Figure 4.12). Leader use for both species varied greatly among

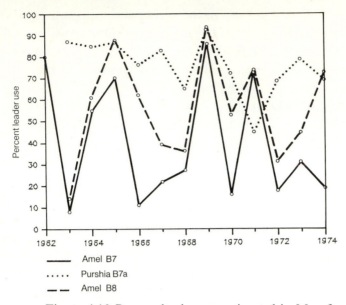

Figure 4.10 Browse leader use estimated in May for *Purshia tridentata* and two samples of *Amelanchier alnifolia* (AMEL) in Spread Creek valley of Grand Teton National Park, 1962–74 (R. P. Wood, unpublished).

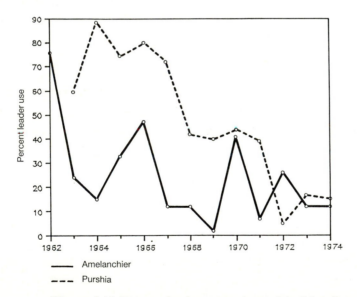

Figure 4.11 Browse leader use estimated in May for *Amelanchier alnifolia* on Lozier Hill and *Purshia tridentata* on Uhl Hill in Grand Teton National Park, 1962–74.

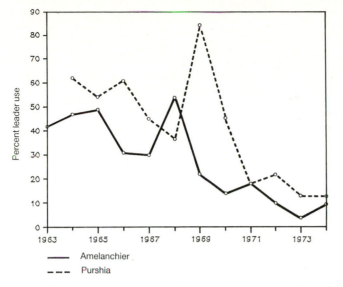

Figure 4.12 Browse leader use estimated in May for *Amelanchier alnifolia* and *Purshia tridentata* on Buffalo Hills sampling units in Grand Teton National Park, 1963–74.

years at Spread Creek, but showed no consistent trends (Figure 4.10). Averaging leader use over all study plots shows a declining trend over the period 1962 to 1974, whereas the fraction of plants that were severely hedged peaks in 1962–63 and again 1970–73 (Figure 4.13).

Although both serviceberry and bitterbrush are preferred by elk, bitterbrush is generally more highly preferred (see review by Nelson & Leege, 1982). Indeed, this appears to be consistent with the browse utilization data from Grand Teton National Park, where mean leader use was 51% for bitterbrush but only 31% for serviceberry. Likewise, 42% of the bitterbrush plants were severely browsed during 1963–74 compared to 35% of the serviceberry plants.

Cole (1969) interprets the declines in browse use over the period 1962–67 covered in his book to be a consequence of 1961–62 being a particularly tough winter with heavy snow, followed by several winters that were more mild. During tough winters less herbaceous forage is available to elk and consequently they use shrubs more vigorously in those years. The continued decline in utilization through 1974 causes one to wonder if this is a correct interpretation, although average use was somewhat higher in 1969 following a heavy snowfall winter (Figure 4.13).

Browse use on Spread Creek fluctuates greatly from year to year (Figure 4.10). R. P. Wood (pers. comm.) hypothesized that this occurs because the transects are located in gullies and may become buried with snow in some winters. However, the data are consistent with Cole (1969), since leader use is positively correlated with maximum depth of snow on the ground in January at Moose, Wyoming ($r = 0.783$, $n = 13$, $P = 0.002$).

Interpretation of some of these patterns is confounded by moose browsing. Moose utilization is particularly important at transect locations on Lozier Hill and Uhl Hill which usually see little winter use by elk (Figure 4.11). In recent years, moose have begun using transects on Spread Creek (Figure 4.10).

Buffalo Hills browse transects are only about 35 m above highway 26/287. Therefore, declining use during 1963–74 may reflect increasing traffic and consequent avoidance of the area by elk, moose and mule deer (Figure 4.12). The decline in browse use on these transects may account for the average overall trend depicted at Figure 4.13.

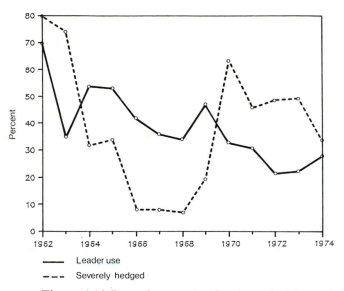

Figure 4.13 Browsing on *Amelanchier alnifolia* and *Purshia tridentata* averaged over all plots in Grand Teton National Park, 1962–74. Plotted is the percent leader use as well as the percentage of plants classified as being severely hedged.

In conclusion, grass and shrub use by elk in Grand Teton National Park varies considerably among sites and among years. Most of this variation can be attributed to (1) cattle grazing within the Park and avoidance of cattle by elk, (2) human disturbance, and (3) weather conditions which alter autumn and spring use patterns – heaviest use of shrubs occurs when snow covers herbaceous forages.

Migration route habitats

Migration from summer to winter ranges may occur during a few days or may span several weeks (Chapter 2). If pushed along by winter storms, migration may occur very quickly with little feeding along the way. Under such circumstances, optimal forage/cover ratios are undoubtedly different from what is optimal on summer range. The value of open forage areas is to provide food (Thomas, 1979), but during migration this may be of less importance. Rather, cover is of utmost importance, especially in areas open to hunting. Human disturbance along migration routes can cause migration routes to shift, and elk migrating through open areas are easier prey for hunters. Thus a key element to migration route habitats is security cover.

Winter and spring ranges

Principal winter ranges for the Jackson elk herd are the National Elk Refuge and surrounding Forest lands, and the upper Gros Ventre valley (Figure 2.3). Other winter ranges on Spread Creek, and along the Buffalo Fork River host variable numbers of elk, but none of these other areas are very large.

National Elk Refuge

The National Elk Refuge is not managed simply as a feedlot, but rather it is managed as an extensive winter range for elk. Although elk have been supplementally fed in all but nine years since 1912, the management objective for the Refuge is to minimize supplemental feeding costs by maximizing the production of winter forage on the Refuge (Wilbrecht, 1983). Substantial numbers (Figure 3.1) of elk use the Refuge for approximately six months out of the year, but they typically free-range for four months and use supplemental feed for slightly more than two months (mean = 76 days, 1960–85).

The US Fish and Wildlife Service maintains vigorous programs to enhance forage production on the Refuge, by prescribed burning, seeding and flood irrigation. As of 1984, 577 ha were flood irrigated, and 1641.3 ha are in seeded grasslands. Current objectives are to burn at least 200 ha annually, and most lands on the Refuge are scheduled for burning on a 4–6 year rotation (J. Wilbrecht, pers. comm.). Burning substantially increases the quantity and quality of forage produced and may reduce the transmission of disease and parasites by destroying elk feces. Each metric ton of increased forage yield produced by prescribed burning costs only $2.65 (Wilbrecht, 1983).

Vegetation types on the National Elk Refuge are categorized in Figure 4.15. Since 1973, forage production has been estimated on each of ten subunits of the

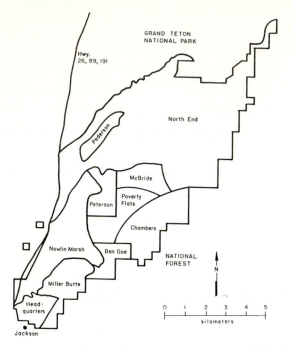

GRAND TETON
NATIONAL PARK

Hwy.
26, 89, 191

North End

Pederson

McBride

Poverty
Flats

Peterson

Chambers

Nowlin Marsh   Ben Goe

NATIONAL
FOREST

N

Miller Butte

Head-
quarters

Jackson

0   1   2   3   4   5
kilometers

Figure 4.14 Map of management areas within the National Elk Refuge near Jackson, Wyoming.

Refuge (Figure 4.14). Estimates entail clipping ten plots on each of 64 transects each fall. Prior to 1983, only herbaceous vegetation was clipped, but since 1983 shrubs have been included in the forage production estimates. Because the transects are not permanent, clipped sites vary from year to year. Consequently, these estimates may show substantial sampling variance (Wilbrecht, 1983). Permanent transects were established in 1986.

Herbaceous forage production on the Refuge (Figure 4.16) increased significantly from 1973 through 1985 ($r_s = 0.648$, $n = 13$, $P < 0.02$). Between 1973 and 1984 forage production declined on the Pederson subunit ($r_s = -0.615$, $n = 12$, $P < 0.05$), but increased on the Ben Goe subunit ($r_s = 0.629$, $n = 12$, $P < 0.05$) and North End ($r_s = 0.915$, $n = 10$ deleting 1983 and 1984, $P < 0.001$; Figure 4.17). The decline in forage production on the Pederson unit can be explained by termination of irrigation on this area since 1980.

The 1984 Annual Narrative Report from the National Elk Refuge suggests that some of the annual variation in forage production is attributable to rainfall during the growing season. Therefore, for each of the subunits I tested the null hypothesis of no correlation between forage production and precipitation at Jackson during and prior to the growing season (January through July). Because 1983 and 1984 included shrub production whereas other years did not

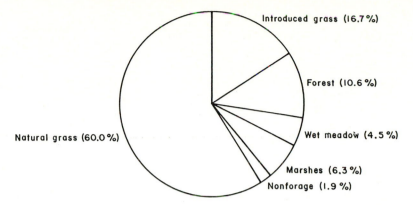

Introduced grass (16.7%)

Forest (10.6%)

Wet meadow (4.5%)

Natural grass (60.0%)

Marshes (6.3%)

Nonforage (1.9%)

Figure 4.15 Broad classification of habitats for the National Elk Refuge (data from Wilbrecht, 1983).

(Appendix I), I eliminated these years for areas on the Refuge where shrub production appeared to contribute to forage production estimates, for example, on the North End and Miller Butte. I found no significant correlations except for Ben Goe where forage production is positively correlated with May precipitation ($r = 0.623$, $n = 12$, $P < 0.05$) and on North End where production is positively correlated with both May precipitation ($r = 0.59$, $n = 10$, $P < 0.05$) and total precipitation for May, June and July ($r = 0.656$, $n = 10$, $P < 0.05$). Forage production and temperature during the growing season were not correlated for any subunits.

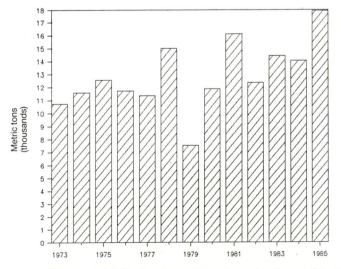

Figure 4.16 Herbaceous forage production estimated for the National Elk Refuge, 1973–85. Intensified irrigation, seeding, and controlled burning have yielded a significant ($P < 0.05$) increase in herbaceous forage production during this period.

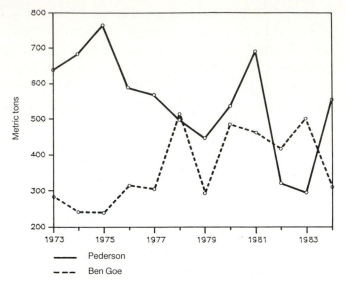

Figure 4.17 Estimated forage production on the Pederson and Ben Goe management areas of the National Elk Refuge. No significant trends existed for forage production estimates on any other management areas of the Refuge.

Forage utilization on the National Elk Refuge in spring from 1980 to 1984 (Table 4.1) is similar to that observed in Grand Teton National Park, albeit during different years. Methodology for estimating utilization is similar: on Park transects, a plant is examined at every third step until a total of 100 plants has been tallied, whereas on the Refuge, 100 plants are inspected but at every other step.

Gros Ventre winter range

Winter range in the Gros Ventre valley is extensive in area (Figure 2.3). Although elk feeding occurs at three State feedgrounds in the Gros Ventre, substantial numbers of elk overwinter in the area each year without supplemental feed.

Cattle grazing in some areas of the Gros Ventre elk winter range exceeds levels prescribed by the US Forest Service. Some riparian areas within the Fish Creek Cattle and Horse Allotment are regularly overgrazed by livestock including the North Fork of Fish Creek, Harness Creek, Lower Squaw Creek, Red Creek, and Beauty Park Creek. Although some of the areas are outside the area currently recognized as elk winter range by the Forest Service, they fall within the boundaries of the area designated by the Forest Service to be elk winter range in 1919, and within the boundaries of elk winter range outlined by the Wyoming Game and Fish Department (see Figure 2.3). Particularly heavy cattle use occurs year after year in the vicinity of the Patrol Cabin feedground; for example, on 19 June 1985, S. Wiseman (US Forest Service, unpublished) reported that cattle had already used 70% of the grass on State lands immediately surrounding the feedground. Because these riparian areas are attractive to cattle, grazing permit holders have difficulty keeping cattle from using these lands.

Over the years, considerable data on shrub and grass use by cattle and elk on the Gros Ventre winter range has been compiled by the Forest Service and the Wyoming Game and Fish Department. Forest Service range inspections are mostly to monitor specific instances of range use by cattle, and have not been conducted in a manner that allows analysis of trend. Table 4.2 summarizes browse utilization by elk and cattle on 29 permanent transects sampled throughout the Gros Ventre winter range annually from 1957 to 1963 (Yorgason, 1963). These data indicate remarkably high levels of browse use, with lowest intensity in mild winters when elk forage over a broader area. Use was heaviest on willow (*Salix* spp.), rubber rabbitbrush, and aspen. Excessive browse utilization (i.e., in excess of 55% leader use) was noted in the Gros Ventre River bottom from Fish Creek to Crystal Creek, the Upper Slide Lake area, and the south end of Sportsman's Ridge.

Wire cages for assessing herbaceous vegetation utilization were monitored in conjunction with the browse transects (Yorgason, 1963). Herbaceous vegetation use exceeded 90% on many areas including above Upper Slide Lake, above

Table 4.1. *Forage utilization on the National Elk Refuge, 1980–84. Number of transects in parentheses.*

Management area	Area, ha	Utilization, %				
		1980	1981	1982	1983	1984
Headquarters	286.2	35	25	60	42 (3)	65 (3)
Nowlin	1059.5	55	50	85	73 (4)	52 (7)
Miller Butte	763.2	25	30	40	38 (2)	21 (2)
Ben Goe	237.7	45	45	50	47 (2)	81 (3)
Peterson	344.9	55	60	70	65 (3)	57 (3)
McBride	215.0	60	55	60	85 (3)	78 (5)
Poverty Flats	481.8	65	70	70	72 (2)	64 (3)
Chambers	743.3	55	45	50	57 (2)	63 (4)
Pederson	263.3	25	25	10	43 (2)	4 (1)
North End	5425.5	25	20	10	16 (5)	16 (3)
Total	9820.6		Weighted mean		35	32

Breakneck Flat, Red Hills, Gros Ventre bottoms–Grey Hills, Alkali Creek and Sportsman's Ridge (Wilbert, 1959). Even though heavy snows accumulated in the valley during mid-winter, elk used a high fraction of the vegetation, consuming it before and after snow accumulation. Very little elk use occurred on the winter range during summer or early fall (Wilbert, 1959). During winter and spring, ridges surrounding the valley were particularly heavily grazed and browsed by elk, for example, above Upper Slide Lake. Data from Yorgason's transects have not been collected since 1973, but elk use of winter range is still heavy. There are presently over twice as many elk on the Gros Ventre feedgrounds than during the early 1960s (see Figure 3.2).

Calving areas

Calving areas are thought to be of particular importance in habitat management for elk (Skovlin, 1982). Elk apparently prefer sagebrush habitats on gentle slopes near forested areas which offer escape cover (Johnson, 1951; Anderson, 1958), usually within 400 m of water. Aspen groves may also be popular calving grounds (Kuck, Hompland & Merrill, 1985; DeByle, 1985c), but it is difficult to generalize because elk will use a wide variety of habitats for calving (Skovlin, 1982). Although some areas are used year after year for calving, snow conditions can alter use of calving areas (Sweeney, 1976).

Known calving areas for the Jackson elk herd (Figure 4.18) include Burro Hill and surrounding areas along the Buffalo Fork (Altmann, 1952; Anderson, 1958), Uhl Hill (Sauer, 1980b), Potholes at the base of Signal Mountain, Ditch Creek, and Sagebrush Flat on Spread Creek (Anderson, 1958). Murie & Murie (1966) record traditional calving areas on grassy slopes above Two Ocean Lake and in meadows surrounding the lake. In the Gros Ventre drainage, calving is common on winter ranges along Dallas Creek, Lightning Creek and Slate Creek (Anderson, 1958), as well as the Haystack Fork–Bear Paw Fork area,

Table 4.2. *Average percent browse use by wildlife and livestock on the Gros Ventre elk winter range. Data are averages from 29 browse use transects located throughout the area (Wilbert, 1959; Yorgason, 1963).*

Species	1957–58	1958–59	1959–60	1960–61	1961–62	1962–63	Mean
Willow	56%	69%	67%	86%	77%	72%	71%
Rabbitbrush	51%	67%	51%	82%	79%	51%	64%
Aspen	44%	69%	50%	66%	49%	48%	54%
Chokecherry	22%	40%	27%	56%	44%	31%	37%
Serviceberry	33%	54%	43%	69%	47%	31%	46%
Mean	41%	60%	48%	71%	59%	47%	54%

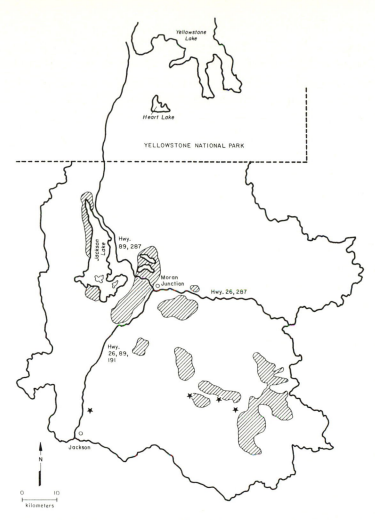

Figure 4.18 Map of known elk calving areas in the Jackson herd unit. A few calving areas are known to exist in southern Yellowstone National Park but exact locations have not been identified. Black stars indicate the locations of winter feedgrounds.

Sohare–North Fork of Fish Creek–Red Creek–Purdy Basin areas, and the Rat Creek–Bacon Creek area (Long *et al.*, 1980).

Consequences of human alterations of elk calving areas are not clearly known, although calving areas are usually considered to be 'crucial' habitats for elk. For example, elk preferred areas with abundant accumulation of downed woody material for calving in the Uinta Mountains of Utah (Winn, 1976). But in western Montana, slash from logging operations can reduce elk use by as much as 50% (Lyon, 1976; Marcum, 1976), although the Montana studies did not focus on calving areas. Elk will readily abandon calving grounds when disturbed (Kuck, Hompland & Merrill, 1985), although there does not appear to be any evidence that disturbance results in abandonment of calves by their dams as postulated by Sauer (1980a). Long *et al.* (1980) recommend avoidance of human activity and cattle grazing on elk calving areas between 1 May and 30 June.

Interactions with other ungulates

Elk are generalist herbivores with highly variable food habits depending upon available forage. Although it is often generalized that elk are grazers, they may forage extensively on browse, particularly in winter. As a consequence of their variable diets, potential exists for competition with several ungulate species that coexist with elk in Jackson Hole. Rigorous demonstration of competitive interaction requires extensive study, and ideally, manipulation of the numbers of both species to document the consequences to population growth rate attributable to variation in density of the competing species. Such investigations typically are not feasible for large ungulates, and usually competition is inferred from similarity of food habits, distributional overlap, and behavioral interactions between species.

In this section I will highlight what is known about interactions between elk and cattle, bighorn sheep, bison, moose, and mule deer. In addition, horses and a few domestic sheep occur within the herd unit, but they are much less numerous and I am unaware of any areas of substantial conflict between elk and either horses or domestic sheep.

Cattle

Heavy livestock grazing can be detrimental to elk range (Nelson, 1982; Lyon & Ward, 1982). It is generally agreed that consequences of competition between elk and cattle are most serious on elk winter–spring range (Stevens, 1966), partly because impacts on vegetation will have most serious consequences when energy balance is especially critical (Nelson & Leege, 1982; Wickstrom *et al.*, 1984). Competitive effects of cattle grazing on elk are often particularly serious on ridge tops and south-facing slopes (Stevens, 1966; Grover & Thompson, 1986).

In addition to the range consequences of cattle grazing, there is known to be direct avoidance by elk of areas being grazed by cattle (Skovlin, Edgerton & Harris, 1968; Mackie, 1985). This was the case in the Gros Ventre study (Long *et al.*, 1980) as shown in Table 4.3; and documented by the fact that radio-collared elk moved out of areas stocked with cattle. In a study of the response of elk to a deferred rest-rotation system for cattle grazing, elk avoided cattle during calving in June and rut in September (Oakley, 1975). This social intolerance of cattle by elk may also influence seasonal movements of elk in the Gros Ventre (Long *et al.*, 1980). When cattle moved into areas used by elk, typically in June, elk often moved to higher elevations where they formed larger groups (Figure 4.19). Later, however, elk moved to higher elevations whether or not they were disturbed by cattle (Figure 4.20).

Because summer ranges on the Jackson herd unit are extensive and forage is not overly exploited by elk, controlled livestock grazing is thought to be compatible. Most livestock grazing on elk summer ranges in Bridger–Teton National Forest is managed on a rest-rotation or deferred grazing schedule. Rest-rotation allows extensive livestock grazing of an area during one summer, but then total rest from livestock use during the next season. Deferred grazing schedules only allow livestock use during prescribed periods of the year. Although most of these grazing management programs are carefully planned, enforcement of grazing plans is poor in some areas within the Forest, particularly in the Breakneck Unit along the upper Gros Ventre River. Forest Service allotment files indicate that the most frequent permit violation is failure to ensure cattle distribution as specified in annual grazing plans. Consequently, some areas may be severely overgrazed whereas other areas receive little or no cattle use.

Although livestock grazing can be detrimental to elk, it actually can be used to enhance elk range if properly regulated. For example, fall grazing by cattle can hasten subsequent spring green-up and improve early spring ranges by removing dead foliage that may slow the growth of plants (Seastedt, 1985; Gordon, 1988), although plant growth proliferates so rapidly in spring that accelerating its appearance is probably of minor value for elk. On the other hand, spring grazing can favor shrubs by reducing the vigor of competing grasses at a critical time in their development (Mueggler, 1972). Thus on winter

Table 4.3. *Percent of time heavy elk and cattle use was recorded when cattle and elk use, respectively, were light or non-existent along the inspection routes (from Long* et al.*, 1980).*

Year	Percent heavy elk use when cattle use was light or non-existent	Percent heavy cattle use when elk use was light or non-existent
1976	80	89
1977	100	100
1978	100	88
1979	93	97

ranges where shrubs provide important winter forage for elk, spring cattle grazing can improve elk habitat.

On the Bridge Creek Wildlife Management Area in northeastern Oregon, cattle were at one time excluded from elk winter range for three years. However, Anderson & Scherzinger (1975) claimed that forage became increasingly 'rank' and of low quality, although no data were collected on forage quality. The nutritional value of the mature forage for elk during winter was presumably improved by cattle grazing in May through mid-June, then no grazing until elk used it (Anderson & Scherzinger, 1975). Similarly, on spring ranges in the Elkhorn Mountains of Montana, elk preferred areas previously grazed by cattle apparently because cattle removed dry vegetation (Grover & Thompson, 1986).

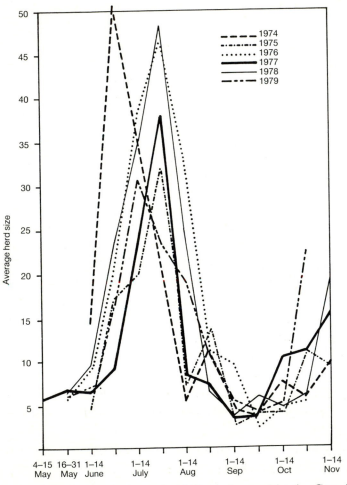

Figure 4.19 Average herd sizes observed in the Gros Ventre Cooperative Elk Study (Long *et al.*, 1980).

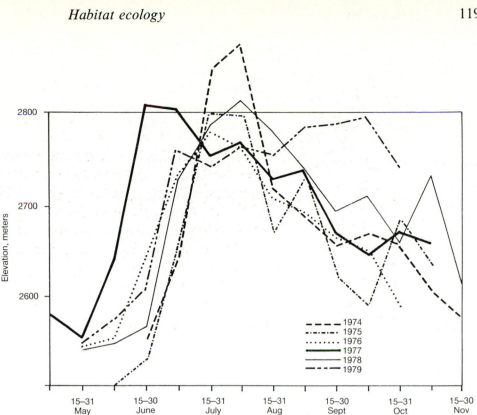

Figure 4.20 Elevation of elk sightings during the Gros Ventre Cooperative Elk Study (Long *et al.*, 1980).

However, spring cattle grazing on elk winter range in the Gros Ventre River valley may not be a good idea, because at least two major forage grasses, bluebunch wheatgrass (*Agropyron spicatum*) and Idaho fescue (*Festuca idahoensis*), are sensitive to early season grazing. Recovery required three years for Idaho fescue and six years for bluebunch wheatgrass when plants were clipped during flowering (Mueggler, 1975). Cattle grazing caused a decrease in Idaho fescue in Jackson Hole (W. B. Jones, 1965). As a consequence of particular sensitivity to early summer grazing, D. Despain (Yellowstone National Park biologist; manuscript) recommends avoiding any grazing on bluebunch wheatgrass before it matures, particularly during late May and June. In portions of the Gros Ventre, bluebunch wheatgrass does not mature until early August.

Elk–livestock interactions were studied for four years in Shoshone National Forest immediately east of the Jackson herd unit (see Figure 9.7) by D. A. Jones (1985). With very heavy November to June elk use of the range, followed by no summer or fall use, no deterioration in range condition was documented, except that sagebrush declined. However, when the same elk use was combined with cattle grazing between June and October, range condition declined.

Few detailed studies exist on the consequence of elk grazing and browsing to plant community structure in Jackson Hole. One exception is the work by W. B. Jones (1965) in which plant species composition was monitored within and outside 32 exclosures in areas where foraging was occurring by elk at some sites and by cattle at others. Plant composition on grazed elk ranges was found to be significantly different from grazed cattle ranges. Elk grazing resulted in decreases in Hood's phlox (*Phlox hoodi*), fringed sagebrush (*Artemisia frigida*), rubber rabbitbrush (*Chrysothamnus nauseosus*), American vetch (*Vicia americana*), tapertip hawksbeard (*Crepis acuminata*), and prairie June grass (*Koeleria cristata*). At the same time, increases were observed for other plant species including rose pussytoes (*Antennaria rosea*), needle-and-thread grass (*Stipa comata*), baldhead sandwort (*Arenaria congesta*), and fleabane (*Erigeron* spp.).

Most importantly, Jones (1965) found that Idaho fescue was reduced on ranges grazed by cattle, yet this important grass was not affected by elk grazing. In addition, cattle grazing caused a decrease in Richardson's geranium (*Geranium richardsonii*), prairie June grass, and Hood's phlox. Only needleleaf sedge (*Carex eleocharis*) and fringed sagebrush increased on cattle range, whereas tapertip hawksbeard did not seem to be affected by cattle grazing.

In summary, competitive interactions between cattle and elk are complex and highly dependent upon details of phenology and plant species composition. Available evidence from the Jackson herd unit indicates that cattle use of an area substantially reduces the value of the area for elk. During summer and fall, elk actively avoid areas occupied by cattle. Cattle use of winter and spring ranges reduces the forage value of these critical seasonal ranges for elk.

Bighorn sheep

Competitive interactions between elk and bighorn sheep in the Jackson herd unit are most likely in the Gros Ventre Mountains. During summer and fall, elk may be seen grazing near bighorn sheep on Sheep Mountain east of the National Elk Refuge. Other areas of potential summer range overlap exist in the Teton Range, the Teton Wilderness and in southern Yellowstone National Park, but it is unlikely that significant competition exists in any of these areas.

Generally, bighorn sheep occupy more rugged terrain and higher elevations than elk, although areas of potential conflict certainly exist. Bighorn sheep and elk winter ranges overlap, both in the Gros Ventre valley and adjacent to the

National Elk Refuge (see Chapter 9). Food habits studies suggest a high potential for diet overlap between the two species (Stelfox, 1976; Barmore, 1986), but interactions between elk and bighorn sheep have not been extensively studied within the Jackson herd unit.

Bison

A small number of bison, less than 25, have lived in Grand Teton National Park since 1969. Since 1980, however, the bison have wintered on the National Elk Refuge and their number has increased to over 100 animals. Food habits of elk and bison overlap extensively although bison more readily forage on sedges (*Carex* spp.) than do elk. Bison recruitment in Yellowstone National Park appears to be suppressed by high elk numbers (Houston, 1982: 180). Nevertheless, numbers of bison on elk summer range in Grand Teton National Park are so small that competitive interactions are minimal.

On the National Elk Refuge, bison management has become a significant management issue. Bison are substantially larger than elk and dominate on the feedgrounds, prompting the Refuge staff to feed bison in a separate area to minimize interactions with elk. Nevertheless, bison are large, expensive to feed, and often cantankerous and dangerous. Therefore, the US Fish and Wildlife Service, the Wyoming Game and Fish Department, and the National Park Service are presently formulating a bison management plan to reduce the bison herd. Reducing the number of bison will have negligible effects on the elk herd, simply because bison numbers are low to begin with.

Moose

Moose occur on both summer and winter ranges occupied by elk in the Jackson herd unit. Moose forage on a diversity of plants, including aquatic plants in summer, and mostly browse during winter (Houston, 1968). Like elk, moose often migrate during summer into higher elevation ranges that are unsuitable in winter because of excessive snow accumulations.

Interactions between elk and moose are not likely to be serious during summer, but on winter ranges, moose may reduce critical browse supplies for elk. Conversely, elk may reduce browse normally used by moose, particularly near feedgrounds in riparian habitats. Also, Martinka (1969) suggested that summer use of riparian willow by elk in Grand Teton National Park may reduce the winter carrying capacity for moose.

On winter ranges, it is common for moose to use hay that has been placed for livestock (see Figure 4.21). However, moose are displaced by elk on feedgrounds, even though the moose may be larger in body size.

Figure 4.21 Moose feeding on hay with horses on the Triangle X Ranch within Grand Teton National Park. Moose are displaced by elk on feedgrounds. Photo by M. S. Boyce, February 1987.

Mule deer

In general, elk tend to be grazers and mule deer tend to be browsers and thereby diet overlap is low. This generalization holds up in many instances (Leslie, 1983), but food habits for both species vary seasonally and there are some times and places where mule deer and elk forage on the same plants, particularly when elk use browse on winter range. Use of browse by elk on the Gros Ventre winter range is extensive (Yorgason, 1963) and this may limit the use of the area by mule deer. Unfortunately, however, no detailed studies on the interaction between mule deer and elk have been conducted in Jackson Hole.

Elk versus aspen

Murie & Murie (1966) wrote 'there is an especially warm spot in our hearts for the quaking aspens that rim our valley and clothe the foothills'. Particularly in autumn, aspens (*Populus tremuloides*) add much color and beauty to the landscape of Jackson Hole, and provide important habitat for many songbirds and other species – including elk (Flack, 1976; DeByle &

Winokur, 1985). It is no wonder that there is considerable concern over the gradual decline of aspen in Jackson Hole.

Much has been written about the consequences of elk browsing on aspen (Krebill, 1972; Beetle, 1974a; Gruell & Loope, 1974; Boyce & Hayden-Wing, 1979). There is no question that elk browsing can substantially impair recruitment of young trees into an aspen stand, particularly when elk are concentrated such as near feedgrounds (Krebill, 1972; DeByle, 1985a). Aspen are heavily browsed on winter range in Jackson Hole, as well as along migration routes, particularly those used during spring migration (Figure 4.22; DeByle, 1985c). Based upon declining condition of aspen in Jackson Hole, Beetle (1974a, 1979) concludes that there are too many elk in Jackson Hole, just as he did about his studies on Big Game Ridge (1962). Emphasizing concern over the loss of deciduous forest habitats, Beetle (1979) pleads that we 'Bring the bluebirds back to Jackson Hole'.

Current condition of aspen in Jackson Hole generally appears to have improved since the early 1970s. Beetle (1974) claims that 'within the influence of feed stations overlapping radii of 20–50 miles, not only tree bark but all available browse is utilized every year'. In contrast, 63% of 484 aspen stands

Figure 4.22 A decadent aspen stand on the southwest slope of Lozier Hill in Grand Teton National Park. Photo by M. S. Boyce.

sampled in Grand Teton National Park showed aspen regeneration with stems taller than 2 m and less than 5 cm diameter breast height ('so that only stands that contained young aspen which had grown beyond most ungulate use were recorded as having aspen reproduction'; Kay, 1985). This represents over 15 times the incidence of aspen stands containing regeneration found in Yellowstone National Park (Kay, 1985).

Aspen are faring less well in the upper Gros Ventre valley where elk and livestock use is more extensive. Kay (1985) reports 36% of 550 stands sampled near roads in 1983 showed successful regeneration, whereas Hart (1986) found only 6% of a more representative sample of stands to contain viable regeneration in 1985.

Elk prefer aspen over other available habitats, at least in some seasons (DeByle, 1985c: 142; Collins & Urness, 1983), although elk populations do not depend upon aspen. For example, DeByle (1985c) notes that good populations of elk occur in some areas like northern Idaho where aspen is a 'minor component of the vegetation complex'. Although most preferred on fall, winter and spring ranges, aspen is used throughout the year. For example, Collins & Urness (1983) found elk to prefer aspen stands over adjacent clear cuts in summer. Aspen appears to be preferred over coniferous forest during all seasons except when thermal or hiding cover is required (DeByle, 1985b).

Although considerable research has attempted to clarify interactions between aspen and elk (Gruell & Loope, 1974; McNamara, 1979; Olmsted, 1979; Weinstein, 1979), many questions remain unanswered. Krebill (1972) predicted that by 1985 aspen stems should have declined approximately 44% in the Gros Ventre River valley if trends he observed remained constant. Indeed, in 1985 the number of stems had declined 39% in the same localities studied by Krebill (Hart, 1986). Mortality of aspen was often caused by pathogenic fungi (primarily *Cenangium singulare* and *Cystospora chrysosperma*) stimulated by browsing damage imposed by elk and other ungulates. Remarkably, however, aspen decreased within exclosures as well as in areas browsed by elk. The cause for declining aspen within exclosures is not understood, although root rot fungus (*Ganoderma applanatum*) may be partly responsible (Hart, 1986).

Gruell & Loope (1974) emphasize the importance of fire for maintaining regeneration in aspen stands. Due to extensive fire suppression in recent decades (Leege, 1968), forest and range fires have been mostly eliminated, or suppressed before they have burned extensive areas. It has been convincingly demonstrated that fires are a major cause of spatial diversity in forest types in the region (Romme & Knight, 1981; Romme 1982), and without fire, aspen stands will be less frequently rejuvenated (Gruell, 1979).

Aspen reproduce mostly by suckering of new shoots from extensive root systems which may cover hundreds of hectares. Burning of aspen stands enhances suckering and new growth in the stand by (1) removing large trees that may compete for water and nutrients with young trees, (2) reducing apical dominance of burned trees, thereby eliminating growth inhibiting auxins produced by the apical bud that suppress growth in lateral shoots or suckers, and (3) releasing nutrients into the soil, thereby stimulating growth conditions for small trees. When 'released' by fire, young aspen trees usually grow very quickly and are able to overshadow competing plants, for example, conifers. In the absence of fire, however, conifers may encroach on aspen stands, particularly when browsed by ungulates (Kay, 1985). Moderate fire tends to suppress many coniferous species whereas it stimulates aspen (Beetle, 1974a).

Gruell & Loope (1974) suggest that it may not be necessary to enforce substantial reductions in elk numbers to sustain aspen in Jackson Hole. Rather, an extensive fire management program offers an alternative solution. In fact, they argue that reducing elk numbers may not be effective in stimulating aspen in the absence of fire.

Aspen rejuvenation on a 160 ha experimental burn on Breakneck Ridge of the upper Gros Ventre elk winter range was evaluated by Bartos & Mueggler (1979, 1981). Although a high intensity burn suppressed sprouting in the first year following fire, aspen suckering increased to nearly double that of a control area during the second and third years after burning (Bartos & Mueggler, 1979).

To be effective, however, burning must be relatively extensive. Bartos & Mueggler (1981) and Hart (1986) observed that the Breakneck Ridge experimental burn was heavily used by wildlife and livestock. Increased use was of such a degree that effective reproduction could not take place even with the increase in suckering stimulated by fire (Bartos & Mueggler, 1981). If more extensive areas are burned, this insular effect will not be so serious and benefits from burning may be realized.

It has not yet been shown that aspen can sustain intensive elk browsing even with ideal fire frequency (D. Despain, pers. comm.). For example, aspen stands burned on the National Elk Refuge have not shown sufficient growth response to outgrow elk browsing (National Elk Refuge, 1984). In the Gros Ventre River valley, aspen sprouts averaged over 1600 per ha, but very few escaped browsing to become established in the overstory. The trees must become large enough to escape being killed by browsing, typically 2 m tall (DeByle, 1985c). This usually requires 6–8 years (Patton & Jones, 1977). Even then, debarking of large aspen can ultimately kill grown trees (Krebill, 1972; Hart, 1986), although this usually only occurs near feedgrounds or during tough winters when elk are hard pressed (Barmore, unpublished). Still, aspen can regenerate with light to moderate browsing, particularly when fire stimulates regrowth (Gruell, 1979).

An alternative to burning to regenerate aspen stands is simply cutting mature trees. Although this does not release nutrients and stimulate sucker growth to the extent that fire can, it may be a useful alternative in areas where it may be difficult to secure an adequate burn. Some aspen stands may be difficult to burn because inadequate fuels exist to maintain a hot fire.

DeByle (1979) proposed that elk and aspen be managed to enforce long-term periodic fluctuations in their populations. For several years, elk numbers could be maintained at high numbers, then culled to low densities. After culling, extensive burning could be used to stimulate growth of aspen which would have a chance to grow because elk numbers were suppressed. Such a plan may be undesirable because it results in substantial variation in elk numbers and harvests, which would be unpopular with some segments of the public. The management scheme which DeByle proposes is straightforward, although details of aspen–elk interaction are too simplistic in his model. Also, such a scheme may not ensure perpetuation of aspen in Jackson Hole. As noted above, aspen stems decreased in the Gros Ventre drainage even when they were not browsed (Hart, 1986). Yet, it is the combined release from browsing and stimulation from burning which may yield desired regeneration of aspen.

Aspen regeneration after leaf blight

Prescribed burning for aspen on Bridger–Teton National Forest has been less than the extent recommended by Gruell & Loope (1974). Since 1983, several aspen stands have been observed regenerating within the Forest without fire, suggesting that there may be reduced need for prescribed burning than was commonly thought. Forest Service personnel postulated that this regeneration may be stimulated by outbreaks of black leaf spot (*Marssonina populi*), also known as aspen leaf blight. Black leaf spot is the most common leaf disease in aspen and is widespread throughout the western United States (Palmer, Ostry & Schipper, 1980). The fungus not only attacks leaves, but also causes a blight of shoots (Boyce, 1948; Mielke, 1957). Twig and branch mortality can occur, particularly after two successive years of infection, and may actually kill the tree. Epidemics are most likely to occur after a wet spring and early summer followed by warm weather (Pinon & Poissonnier, 1975; Hinds, 1985).

If recent outbreaks of black leaf spot in Jackson Hole stimulated regeneration of aspen, this may help to counteract the consequences of fire suppression. This could occur if the trees, or at least the apical buds, were killed by the disease, thus reducing apical dominance, stimulating growth of lateral shoots and root suckers, and thereby regenerating stands. Older, less vigorous aspen clones have a dramatic amount of bud and branch death subsequent to a leaf blight epidemic (Harniss & Nelson, 1984). The disease apparently kills buds rather than stems, and buds that do survive produce vigorous shoots with large leaves. The suggestion that regeneration may be stimulated by blight may have originated with Mielke's (1957) statement that 'In some cases twigs and branches are killed and in others the entire tree may die except for some of the roots. Sprouting from these remaining live roots usually perpetuates the stand to some degree.'

Regeneration of blight-infested aspen stands might not be as great as might occur after burning because nutrients are not released as occurs after fire. Nevertheless, black leaf spot might offer hope that aspen stands will be revitalized in Jackson Hole despite continued fire suppression. An extensive leaf blight epidemic in Jackson Hole and surrounding areas during the summers of 1981 and 1982 (Harniss & Nelson, 1984) was suggested as one cause for regenerating some stands on Bridger–Teton National Forest (Floyd Gordon, pers. comm.).

It appears, however, that black leaf spot seldom causes substantial mortality, and clones vary widely in susceptibility to the fungus (Hinds, 1985). Even when infections are heavy, impacts are most severe on smaller trees, apparently because microclimates for growth of the fungus are less favorable in crowns of taller trees. In most years, infection repeats only in the lower crown of trees (Hinds, 1985). Therefore, it is improbable that apical dominance could be interrupted and regeneration is unlikely to be stimulated by infestations of black leaf spot.

Although *Marssonina* can be common among aspen suckers and small trees, a high incidence of insect- and disease-induced injury is normal in aspen sucker stands (Perala, 1984). Although growth by infected suckers is diminished, 'productivity of these stands should meet or exceed historical expectations'. Perala suggests that insects and disease may be beneficial by promoting normal self-thinning and maturation in aspen stands.

Another possible explanation for the apparent improving status of aspen in Jackson Hole is that root rot disease may be killing large trees and thereby enhancing regeneration of some stands (Dr J. Hart, pers. comm.). But this cannot be considered an alternative for reestablishing the role of fire in regenerating aspen stands, as proposed by Gruell & Loope (1974). Benefits from burning aspen are much broader than simply regeneration of aspen (Gruell, 1980a, b). A variety of wildlife species are benefited, and encroachment on aspen stands by conifers is reduced (DeByle, 1985b).

Summary

1. In remote areas of the Teton Wilderness and southern Yellowstone National Park, elk summer on mountain ridge herblands. In areas frequented by humans, elk usually remain close to forests which provide security cover. Security cover is also a key component of migration habitats.

2. Elk numbers in Grand Teton National Park increased markedly after the expansion of the Park in 1950, apparently as a consequence of removal of livestock from west of the Snake River.

3. Intensified range management on the National Elk Refuge has increased forage production for wintering elk in recent years. Forage production on the Refuge is enhanced by burning, irrigation, and seeding.

4. Cattle compete with elk for forage and elk actively avoid areas grazed by cattle. Combined elk and cattle use of winter ranges in the upper Gros Ventre River drainage is very heavy despite its designation as crucial big game winter range. Competitive interactions between elk and native ungulates do not appear to be serious; however, these interactions have not been the focus of any studies in the Jackson elk herd.

5. Aspen recruitment is severely limited by elk browsing in the vicinity of winter feedgrounds, particularly in the upper Gros Ventre drainage. Prescribed burning stimulates aspen suckering and growth, and may allow some browsed stands to regenerate.

Chapter 5.

Winter feeding

Supplemental feeding of elk in winter is fundamental to the management of the Jackson elk herd. It is the National Elk Refuge management programs and the state and federal winter feeding which make the Jackson elk herd unique. Generally winter feeding is discouraged in wildlife management, but for the Jackson elk herd, justification appears sound because agricultural activity and the town of Jackson have displaced elk from previously occupied winter range. The National Elk Refuge was originally established to ensure range for wintering elk to discourage them from exploiting private hay stacks in the valley (Madsen, 1985). However, it quickly became apparent that forage resources on the Refuge were inadequate to satisfy the large elk herd and winter feeding became a regular program.

In this chapter I survey the winter feeding program for the Jackson elk herd, beginning with a description of the procedures and administration of the feeding program, and the unusual Boy Scout antler auction which provides supplemental funds for the feeding program. Next I discuss factors which determine the optimal timing of feeding and controversies regarding the type and quantity of supplemental feed. Then, to evaluate the desirability of the feeding program, I review evidence pro and con on supplemental feeding, including such matters as disease incidence among feedground elk. This information is integrated into a discussion on the desirability of an additional feedground which has been championed recently by guides and outfitters in the area, and finally into concluding remarks on the ultimate question: 'Is supplemental feeding necessary?'

The National Elk Refuge is currently administered by the Fish and Wildlife Service. Originally covering 11.3 km² when established in 1912, acquisition of additional lands has expanded the Refuge to 98 km² (Wilbrecht, 1983). As discussed in Chapter 4, management priorities have shifted in recent years so that the Refuge is now managed as an elk winter range. Extensive forage production is maintained on the Refuge through burning, seeding and irrigation. This is particularly important for mild winters when elk are able to access forage. Indeed, during nine mild winters, for example, 1976–77 and 1980–81, no supplemental feeding was required, yet extensive use of the Refuge was made by wintering elk.

Elk are also regularly fed at three Gros Ventre feedgrounds by the Wyoming Game and Fish Department. Prior to 1956, feeding at these areas was primarily on an emergency basis (Anderson, 1958), with major feeding only in severe winters. By 1960, elk were fed in the Gros Ventre every year, except for the mild winters of 1976–77 and 1980–81. In addition to the Gros Ventre feedgrounds, occasional emergency feedings have been provided in the past for elk at Burro Hill near the Buffalo Fork and at Elk Ranch Reservoir adjacent to Uhl Hill. Such emergency feedings are to secure the survival of elk when access to natural forage is denied by heavy snow, and to appease the public.

Feeding operations

In early years, loose hay was harvested on Refuge lands and fed to the elk. Beginning in the mid-1920s additional hay was purchased from ranch-

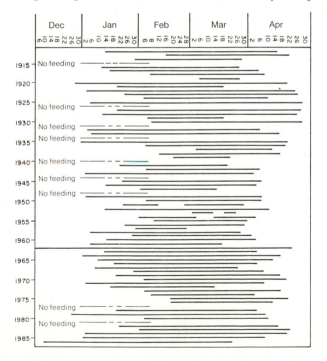

Figure 5.1 Timing of feeding at the National Elk Refuge, 1912–85 (updated from Robbins, Redfearn & Stone, 1982).

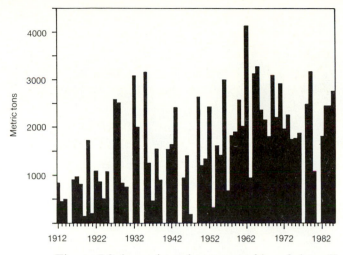

Figure 5.2 Annual total tonnage of hay fed to elk at the National Elk Refuge, 1912–85. To calculate totals for years when pelleted alfalfa was fed, 4 tons of pelleted alfalfa is considered equivalent to 5 tons of baled hay (Thorne & Butler, 1976; Smith & Robbins, 1984).

ers to ensure adequate stocks to feed the elk herd. Beginning in 1938 baled hay was fed, and since 1975 pelleted hay has been fed. The length of time that the elk were fed and the amount of hay fed has varied from year to year (Figures 5.1 and 5.2).

In Figure 5.2 the tonnage of loose and baled hay fed to the elk were not adjusted, but during years when pelleted or cubed hay was provided, I used the conversion that baled hay was equivalent to four-fifths the same quantity of pellets or cubes. Eight hundred tons of pelleted hay are considered to be nutritionally equivalent to approximately 1000 tons of baled hay (Thorne & Butler, 1976; Smith & Robbins, 1984).

Between 1952 and 1974, an agreement with the National Park Service allowed cultivation and harvest of hay on 527 ha in the southeastern portion of Grand Teton National Park. The US Fish and Wildlife Service raised the hay and the Wyoming Game and Fish Department contracted to harvest and store the hay on the National Elk Refuge (Robbins, Redfearn & Stone, 1982). The National Park Service viewed the operation as inconsistent with Park policy and preferred that the Refuge obtain hay elsewhere. Since 1974 baled or pelleted hay has been purchased from private ranchers in Wyoming and eastern Idaho.

Elk wintering on the National Elk Refuge currently are fed in three or four groups. Usually one group is near the Refuge's sleigh ride visitor center north of Miller Butte, one near the maintenance shops at the south end of the Refuge, a

Figure 5.3 Elk feeding behind feed wagon on National Elk Refuge, February 1985. Pellets are in the snow between tracks of the wagon.

group in the Poverty Flats area, and another in the McBride region (Figure 4.16). Pelleted hay is distributed on the snow from a large 20-ton truck or from gravity-dispensing feed wagons pulled behind Caterpillar tractors (Figure 5.3). Only three employees working two to three hours are required to feed all four groups of elk each day.

On the National Elk Refuge, feeders remain on the feedground 30–45 minutes after feeding to observe elk to determine whether more elk are arriving from outlying areas, and to assess whether elk appear content or if it is necessary to provide additional feed. If crowding on feed lines is observed, additional feed is distributed in that location.

Elk on the Gros Ventre feedgrounds are fed baled hay which is distributed by horse-drawn sleds. The Wyoming Game and Fish Department employs seasonal employees who use snowmobiles to access the area during winter. Hay stores are maintained at each of the feedgrounds. Tonnage and costs of hay for State feedground operations in the Jackson herd unit are summarized in Table 5.1.

It is expensive to feed several thousand elk for 2–3 months. Expenses vary substantially depending upon the number of elk being fed, the length of time that feeding takes place, and the cost of feed. Since the early 1950s, costs for supplemental feed on the Refuge have been shared equally by the US Fish and Wildlife Service and the Wyoming Game and Fish Department. Annual costs for feed, labor and equipment can exceed $250000 (Wilbrecht, 1983), and average over $2650 per day that feeding takes place (Madsen, 1985).

Antler sales for feed

Prior to 1957, antlers cast by bull elk were collected on the National Elk Refuge by employees and given to tourists visiting the Refuge. After that year the Jackson District of the Boy Scouts began collecting antlers under permit from the Wyoming State Game and Fish Department and sold them to various tourist facilities in Jackson and to the town of Jackson and Rotary Club

Table 5.1. *Wyoming Game and Fish Department elk feeding programs on Gros Ventre feedgrounds, winters 1966–67 to 1984–85.*

	Alkali feedground			Fish Creek			Patrol Cabin		
Year	No. days fed	Metric tons hay	Total costs*	No. days fed	Metric tons hay	Total costs*	No. days fed	Metric tons hay	Total costs*
1967	46	297.3	$6979	48	218.2	$4798	89	149.1	$5817
1968	83	310.9	$7123	82	220.9	$5117	85	130.9	$5395
1969	91	340.9	$7464	91	253.6	$5514	90	170.0	$8105
1970	98	288.2	$9414	83	274.5	$10486	89	130.9	$5822
1971	100	330.9	$12300	101	300.0	$11501	100	225.5	$8831
1972	71	147.3	$6656	71	190.0	$8076	69	144.5	$7071
1973	91	230.0	$9659	87	213.6	$8012	88	169.1	$8174
1974	99	200.0	$8136	101	255.5	$9820	98	198.2	$10381
1975	111	230.9	$7352	119	276.4	$15968	119	148.2	$8320
1976	106	427.3	$20741	101	402.7	$19111	106	no data	$8754
1977	0	0.0	$416	0	0.0	$0	0	0.0	$972
1978	94	154.5	$16306	94	267.3	$24371	94	177.3	$17882
1979	100	246.4	$23668	103	225.5	$22337	96	205.5	$20352
1980	91	168.2	$13682	90	185.5	$15055	90	215.5	$16397
1981	0	0.0	$0	0	0.0	$0	0	0.0	$0
1982	124	318.2	$28649	109	328.2	$29239	114	212.7	$18947
1983	94	126.4	$13469	110	257.3	$25155	90	114.5	$11749
1984	109	214.5	$20515	109	393.6	$36206	88	60.0	$7308
1985	77	66.4	$7817	90	240.0	$25332	64	69.1	$8290
Mean	83	215.7	$11597	84	237.0	$14531	83	132.7	$9398

* Costs include mileage, hay costs and transportation, expenses for feeding horse teams used for distribution of hay, permanent and seasonal employees, equipment, maintenance, and supplies for feeding program.

to construct the four antler arches in the Jackson town square. During these early years antlers were sold for about $1 per kg or less.

In 1966 the Refuge established a Special Use Permit (SUP) to better control antler pickup and to encourage the Scouts to assist with trash clean up in addition to antler pickup. In 1968 an additional provision was added to the SUP requiring the Scouts to dispose of antlers through a public auction. This was to assure accountability for antlers and to have a more fair availability to the public due to increasing demand. At this time, average price for antlers was about $2.25 per kg.

In the 1973 auction, the average price went over $11 per kg and the Scouts collected $19616. In 1975 the SUP was amended to provide a return to the Federal treasury (Refuge Revenue Sharing Fund) of 25% of the auction proceeds with the bulk being retained by the Scouts. At this time it was established and included in the SUP that the value of the Scouts pickup and auction 'costs' were $7500 and they would be entitled to this amount before any payment was made to the Revenue Sharing Fund. In 1977, for example, only 706 kg were auctioned garnering the Scouts $7400 with none remaining for the government. In 1978, the auction produced $47600, of which $37600 was retained by the Scouts and $10000 went into the Revenue Sharing Fund.

Many of the antlers are sold for crafts and jewelry with German buyers competing in some years. Also, buyers from Korea and Japan purchase antlers which are subsequently ground for use in medicines and as a supposed aphrodisiac (Potter, 1982). Because of increasing value of the antlers and the lucrative 'fund raising' project of the Boy Scout District, the Refuge Manager came under increasing pressure from Girl Scouts and other nonprofit groups to 'get into the act'.

Negotiations with local Scout leaders convinced them that some of this income should be returned to the elk management program, and beginning in 1979, the Scouts established a 'feed fund' with dollars remaining after taking $7500, paying 25% to the Revenue Sharing Fund and retaining an additional 5%. In 1980 the permit fee (Revenue Sharing Fund) share was reduced to 5%. Details of antler pickup and feed fund are presented in Table 5.2. From 1979 through 1986 the Scouts have purchased $225539 of feed and donated it to the National Elk Refuge and Wyoming Game and Fish Department for supplemental feeding of elk on the Refuge. This amounts to over 10% of the total Refuge feeding costs during this period.

When to feed?

Initiation of feeding on the National Elk Refuge has varied from year to year (Figure 5.1). The decision as to when to begin feeding currently is based upon the experience of Refuge personnel and the Wyoming Game and Fish

Department area biologist (Robbins, Redfearn & Stone, 1982). Beginning in late November or early December, snow conditions and the distribution of elk are surveyed and mapped. The following factors are considered in deciding when to begin feeding: (1) number of elk on the Refuge, (2) snow depths in excess of 32 cm, or less if there has been sufficient crusting or packing (Madsen, 1985), (3) apparent condition of the elk, (4) extent to which the elk must paw to gain access to forage, (5) degree of forage utilization, and (6) amount of natural forage available.

On State feedgrounds, similar criteria are used for timing the initiation of feeding. In addition, public pressure to initiate feeding may also be a consideration. In January 1986, ranchers on the Gros Ventre encouraged the initiation of feeding at the Gros Ventre feedgrounds because elk were using hay stocks maintained for private livestock.

An alternative approach is being developed by N. T. Hobbs (Colorado Division of Wildlife, pers. comm.) for timing the initiation of feeding for mule deer (*Odocoileus hemionus*) and elk in Colorado. Hobbs has developed a model

Table 5.2. *Revenues from sale of elk antlers by Boy Scouts.*

Year	Antler mass, kg	Mean price bundled antlers per kg	Total antler sales	Boy Scout income	Funds to US Fish & Wildlife Service	Revenue for feeding program
1968			$3000	$3000		
1969			$4000	$4000		
1970			$3000	$3000		
1971			$3400	$3000		
1972			$4600	$4600		
1973			$19600	$19600		
1974	1589.1	$3.65	$5800	$5800		
1975	2291.4		$13000	$11700	$1300	
1976	1859.3	$8.84	$18000	$15400	$2600	
1977	705.9	$10.48	$7400	$7400		
1978	3574.7	$13.26	$47600	$37600	$10000	
1979	3978.7	$13.26	$51500	$9000	$11000	$31500
1980	3810.9	$13.26	$50000	$9500	$2000	$34500
1981	1081.4	$9.61	$10400	$7600	$145	$2600
1982	3470.1	$13.92	$50430	$9539	$2146	$38744
1983	2103.6	$12.49	$34715	$8793	$1361	$24561
1984	3135.7	$13.28	$44083	$9237	$1829	$33017
1985	3277.3	$13.92	$51378	$9584	$2194	$39601
1986	2465.2	$14.10	$30832	$8608	$1167	$21058
Total						$225581

of energetics with the following simple inputs: (1) snow depth on winter range, (2) number of days with wind chill below -5 °F., (3) number of nights with wind chill below -5 °F. The model then simulates energy intakes and expenditures to predict energy reserves and body mass changes for an average animal in the herd to suggest when feeding should begin to ensure survival for most females (see Swift, 1983). Availability of natural forage is presumed to be directly related to snow depth. Currently, the model has only been refined for deer and will require additional information on elk body fat and energetics to be extended with reliability.

Would development of such a model be useful for timing the initiation of feeding for the Refuge and State feedgrounds? Given some experience, Refuge and State personnel probably integrate in their minds a number of variables very efficiently to time feeding so as to minimize losses of elk. But if change in personnel should occur, or unusual winter weather arise, input from such a timing model could be very helpful. Timing of the initiation of feeding is frequently a source of controversy between local interest groups and agencies. A carefully designed multivariate modeling approach might be a useful tool for justifying feeding schedules to the public.

How much and what to feed

Given that the objective of the winter feeding program is to ensure survival and vigor of overwintering elk, a critical issue is determining an adequate and economical ration. In addition, it is important that adequate stocks of feed be on hand to secure survival of elk even during a severe winter.

The amount of supplemental feed and the length of time it was fed to elk each winter has been recorded since 1912. Unfortunately, however, the per capita ration fed is uncertain prior to 1983 because the only indications of the number of elk fed are the highest minimum counts of elk on the Refuge. These counts may not reflect average elk use of the Refuge, and in fact, in some years the average number of elk on feed can be substantially below the maximum counts (B. Smith, pers. comm.). Therefore, extrapolations to per capita feeding rates are minimum values and considerable variability is unaccounted for. Statistical methodology for instantaneous counts is appropriate here, and I have summarized relevant models in Appendix J.

During years of high elk numbers, feeding rations were significantly lower than when elk numbers were low, both on the National Elk Refuge and on the Gros Ventre feedgrounds. For example, on the National Elk Refuge since 1927, the recommended ration of 3.4 kg of pellets or 4.25 kg of baled hay (Thorne & Butler, 1976) was not fed during any of 11 years that elk numbers exceeded 8000 animals. However, in 14 of the 36 years when 8000 or fewer elk wintered on the Refuge, rations were provided that were greater than Thorne & Butler's (1976) recommended ration (Figure 5.4). There is a very low probability that this pattern occurs by chance:

$$\Pr[X=0] = \begin{bmatrix} 11 \\ 0 \end{bmatrix} \left(\frac{14}{36}\right)^{0} \left(\frac{22}{36}\right)^{11} = \left(\frac{22}{36}\right)^{11} = 0.004 \qquad (5.1)$$

At least during early years of feeding, this pattern may reflect that when elk numbers were large, inadequate feed was available to maintain high rations throughout the feeding period, particularly during severe winters. However, during the past 20 years stocks of feed have never been exhausted.

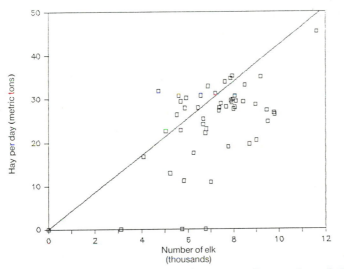

Figure 5.4 Hay fed per day versus the number of elk on feed at the National Elk Refuge, 1927–85. The line represents a daily ration of 3.4 kg of alfalfa pellets or 4.25 kg of baled hay. Smaller daily rations were provided during years when elk numbers exceeded 8000.

In winter 1970–71, the National Elk Refuge began investigations into alternative types of feed for the elk, partly in response to the scheduled 1972 termination of the hay sales agreement with Grand Teton National Park (Robbins & Wilbrecht, 1979). It was decided to begin using pelletized hay because (1) it contains more leaves than baled hay and is therefore more nutritious, (2) it can be fed with less manpower by mechanized equipment, (3) it is cheaper to transport and handle than baled hay, and (4) it was thought that there was less wastage. Although pelleted hay costs more than baled hay, reduced staff and time required to distribute pelleted hay makes it more economical than baled hay (Robbins & Wilbrecht, 1979). Beginning in 1970, pelleted hay was fed on a trial basis to penned elk, and in 1972 the first pelleted hay was fed to a group of elk on the National Elk Refuge (Robbins &

Wilbrecht, 1979). Gradually, feeding shifted until only compressed alfalfa, either pellets or cubes, have been used since 1979.

The Wyoming Game and Fish Department (Thorne & Butler, 1976) and the US Fish and Wildlife Service (Smith & Robbins, 1984; Oldemeyer, Robbins & Smith, unpublished) have investigated the pellet feeding program. At least part of the reason for the Thorne & Butler study was concern that pelleted hay may not offer the roughage value of baled hay and that pelleted hay may consequently be nutritionally inadequate for elk (E. T. Thorne, pers. comm.). The two studies differed in design, but both studies supported the shift to pelleted hay, and did not suggest serious nutritional difficulties associated with feeding pellets. Only one feeding trial was conducted with cubes (Thorne & Butler, 1976) which is inadequate to offer adequate comparison with pelleted or baled hay. Therefore, I will not discuss results from the cubed hay feeding trial.

Wyoming Game and Fish Department elk feeding studies (Thorne & Butler, 1976) were conducted at the Sybille Research Facility northeast of Laramie. Elk were separated by sex and age, and kept in pens which contained no natural forage. Conditions were carefully controlled, but consequently, different from conditions experienced by elk on the Refuge. In contrast, Fish and Wildlife Service studies (Smith & Robbins, 1984; Oldemeyer, Robbins & Smith, unpublished) were conducted in pens on the Refuge. Natural forage existed in the pens, and sex and age composition of penned animals was comparable to that of the wintering elk herd on the Refuge. Although the Sybille studies were not confounded by unknown intake of natural forage, the Refuge studies more closely represented conditions which elk experience on the Refuge.

When offered pellets, cubes or baled hay, all types of feed were taken readily by elk. Both studies found that elk performed better on pelleted hay than on baled hay because less waste occurred with pellets and pellets were more nutritious. Pelleted hay passed through the digestive tract of elk more rapidly, and was slightly less digestible than baled hay (Thorne & Butler, 1976). Loss of 'fines' (i.e., small particles) from pelleted hay appears to me to be difficult to estimate accurately, but both studies concurred that such losses were lower for pellets than was wastage for baled hay.

I have summarized the results of feeding trials at the National Elk Refuge and at the Sybille Research Station in Figure 5.5. Results are pooled within years for elk of the same age/sex group at the same feeding ration. Results are standardized for mean body mass of elk at the beginning of the trial, and standardized for a 100 day feeding trial. Even though native forage was available for elk in pens on the National Elk Refuge, body mass dynamics at various rations did not differ significantly from those observed at Sybille. Possibly greater thermal stress and activity of elk in the National Elk Refuge pens offset available native forage in the pens.

Least squares regression of body mass change versus feeding ration yields a daily maintenance ration of 1.7 kg pelleted hay per 100 kg body mass. For cow elk averaging 216 kg body mass (Table 5.3), this entails a daily ration of 3.7 kg of pelleted hay. Calves usually gained weight during feeding trials because of their lower daily food requirements. Maintenance ration for an average 112 kg calf elk (Table 5.3) is only 1.9 kg of pelleted hay per day. Given the sex/age composition of elk on the Refuge in 1986 (Appendix C), mean weight of elk in the Refuge herd is 199 kg, indicating a daily maintenance ration of 3.4 kg of pelleted hay.

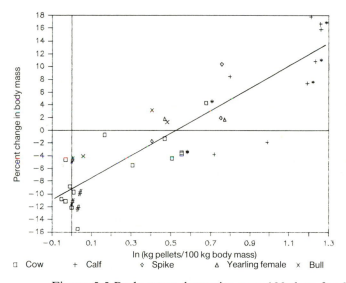

Figure 5.5 Body mass dynamics over 100 days for 345 elk on pelleted hay feeding trials during winters 1972–82 at the National Elk Refuge (Smith & Robbins, 1984; Oldemeyer, Robbins & Smith, unpublished) and at the Sybille Wildlife Research Facility (indicated by * right of symbol; Thorne & Butler, 1976). The four trials marked with # were fed double rations every other day at the National Elk Refuge. Least squares regression line is (% change in body mass/100 da) = 17.4 (ln [ration/100 kg body mass]) − 9.27 ($r = 0.872$, $n = 32$, $P < 0.001$).

These are average rations for maintaining body mass during winter feeding of pelleted hay. However, Oldemeyer, Robbins & Smith (unpublished) note that maintaining weights may not be a necessary objective for the feeding program because seasonal weight fluctuations are normal amongst ungulates in winter (Price & White, 1985). Therefore, these average rations may be considered conservatively high, and lower rations may often be adequate.

In addition to feeding trials on the Refuge, Smith & Robbins (1984) estimated maintenance requirements based upon nutritive quality of pellets. They found 16% crude protein in samples of alfalfa pellets, which is virtually identical with findings of Thorne & Butler (1976). Based upon crude protein requirements for a 260 kg cow elk (one big cow!), they follow Nelson & Leege's (1982) model

$$\text{daily consumption} = 163.3/(260) \; (P_c/100)(D) \tag{5.2}$$

to calculate a daily maintenance ration of 3 kg of pelleted hay. In equation 5.2, the daily digestible protein requirement for a 260 kg elk is 163.3 g, P_c is the percentage of crude protein required in the diet, and D is the average digestion coefficient for diet crude protein. This amounts to only 1.15% of body mass or 44% less than determined by Thorne & Butler (1976). Others have estimated

Table 5.3. *Whole body weights (kg) for elk killed in Grand Teton National Park, November 1973 and 1974 (data from Thorne & Butler, 1976).*

Age and sex class	November 1973	November 1974	Pooled 1973 and 1974
Male calves			
Mean	119.5	108.6	113.6
Standard deviation	17.4	23.4	20.5
n	4	5	9
Female calves			
Mean	110.0	109.5	110.0
Standard deviation	6.9	22.2	18.4
n	3	7	10
All calves			
Mean	115.4	109.5	112.2
Standard deviation	13.8	21.6	19.3
n	7	12	19
Yearling females			
Mean	181.0	133.9	165.2
Standard deviation	32.0	0.0	35.3
n	2	1	3
Spikes (yearling males)			
Mean	174.2	191.9	186.0
Standard deviation	41.6	20.2	26.0
n	2	4	6
Adult cows			
Mean	215.8	216.7	216.3
Standard deviation	26.1	18.3	22.1
n	13	13	26
Mature bulls			
Mean	217.2	256.1	246.2
Standard deviation	0.0	45.8	42.1
n	1	3	4

similar maintenance requirements, using different models (Oldemeyer, Robbins & Smith, unpublished). For a 200 kg elk, Hobbs *et al.* (1982) estimated energy requirements equivalent to a daily ration of 3.6 kg of alfalfa pellets. To maintain a positive nitrogen balance, Mould & Robbins (1981) calculate that a 230 kg elk will require slightly over 3 kg of alfalfa pellets per day.

Estimates of maintenance ration determined directly from feeding trials on captive elk seem more defensible than those extrapolated from presumed crude protein requirements and digestibility ($>60\%$). Elk diet requirements are clearly more complex and include more variables than simply crude protein and digestibility (see, e.g., Hobbs & Swift, 1985), and the results of feeding trials (Figure 5.5) integrate many unknowns not considered at equation 5.2.

Smith & Robbins (1984) emphasize that constant daily rations for supplemental feeding of elk is not reasonable. Instead, rations should be adjusted depending upon a variety of factors, including temperature, snow conditions, and availability of natural forage. On State feedgrounds, another consideration is the need to distract elk from private haystacks. Unfortunately, at this time no detailed model has been developed which can predict the consequences of these confounding variables.

Thorne & Butler (1976) recommend that daily rations of 4.35 kg of alfalfa pellets be provided to elk on feedgrounds. This is higher than maintenance requirements from the penned study results to account for greater activity and therefore greater energetic demands for feedground elk. These recommendations are designed to maintain body mass in elk on feed, and a principal motivation appears to be the results of Thorne, Dean & Hepworth (1976) showing that weight losses for cow elk greater than 3% between calving and parturition can reduce elk calf survival.

However, there is a flaw in this justification for Refuge elk because elk always go off supplemental feed in March or April (Figure 5.1) and use natural forage for the final 6–8 weeks of pregnancy. Pen studies by Thorne, Dean & Hepworth (1976) maintained cows on experimental rations until parturition. Cow elk well fed during the last weeks of pregnancy produced healthy calves even though they may have lost up to 11.1% body weight during the first two trimesters of pregnancy (Thorne, 1981). Feeding may play a greater role ensuring elk reproduction at State feedgrounds where feeding is usually longer in duration.

Similar results are reported by Oldemeyer, Robbins & Smith (unpublished) who found that calf birth weights were not influenced by the weight dynamics of the cow during feeding trials, even when cows lost as much as an average of 8.8% body weight. They attribute this to unlimited forage availability once green-up occurs in March or April and subsequent recovery of the cow during the last trimester of pregnancy. Although weight of the calf clearly influences survival (Thorne, Dean & Hepworth, 1976; Clutton-Brock, Albon & Guinness, 1982; Oldemeyer, Robbins & Smith, unpublished), it appears that cow weight loss of 3% suggested by Thorne, Dean & Hepworth (1976) was overly conservative. When abundant forage is available during the last trimester of pregnancy, this value may be closer to 11–12%.

Inspecting Figure 5.2, it is apparent that feeding levels on the National Elk Refuge have averaged higher during the past 25–30 years than during early years of the feeding program. Indeed, there is a significant positive trend in the total tonnage of hay provided over the period 1912–85 ($r_s = 0.293$, $n = 52$, $P < 0.05$) and likewise the mean quantity of hay fed per day increases over this period ($r_s = 0.417$, $n = 52$, $P < 0.01$). During years prior to 1960 when there was greater year-to-year variability in total quantities of hay fed, there exists a significant positive correlation between tonnage of hay fed during a winter and recruitment rates the following year ($r_s = 0.893$, $n = 7$, $P < 0.01$), consistent with Thorne, Dean & Hepworth's (1976) results that the condition of the dam during pregnancy influences the survival of calves. However, during years since 1960 hay rations have generally been sustained at high levels except during mild winters. Consequently, spanning the entire period 1927–85, we see a positive, but statistically insignificant correlation between hay fed and recruitment rates the following year ($r_s = 0.088$, $n = 35$, $P > 0.1$).

Is it justified that feeding not be instituted during mild winters? Apparently so, because the calf/cow ratios from classification counts in the year following a no-feeding year are not significantly different ($P > 0.1$) from those occurring during an average winter when elk were being fed. In fact, these two means are remarkably similar: for years after no supplemental feed, mean calves/100 cows $= 29.65 \pm 2.68$ SE ($n = 4$); for years following feeding, mean calves/100 cows $= 30.28 \pm 1.14$ SE ($n = 36$).

There do not appear to be negative consequences to feeding pelleted hay although a few cases of bloat have been observed when elk first gained access to large quantities of pellets. To remedy this, Smith & Robbins (1984) recommend gradually increasing the rations over a 7–10 day period to allow for adjustments in rumen microbes.

Competition among different age and sex classes at the feed lines can be severe. Bulls typically dominate, and calves invariably do least well in competitive interactions with older-aged individuals. This is readily alleviated by increasing rations (Thorne & Butler, 1976) and by distributing pellets in an S-shaped pattern over a large enough area that no individual has difficulty gaining access to feed (Smith & Robbins, 1984).

Costs of pellets are 21% higher than for baled hay, but due to greater nutritional quality of pellets, 15–20% fewer pellets need be fed. Assessing which form of hay is more economical requires consideration of loading, equipment, transportation, personnel required for feeding, storage and unloading costs. Calculations by Thorne & Butler (1976) indicated that pellets may not be economical on State feedgrounds, but for the National Elk Refuge where personnel requirements have been reduced, and where equipment is in place, it is clearly more economical to feed pellets than baled hay (Robbins & Wilbrecht, 1979). Improved equipment for mechanical distribution of baled hay has become available in recent years which might improve economics of feeding baled hay. But since an expensive capital investment in equipment for handling alfalfa pellets already exists at the Refuge, this seems to be a moot point.

Advantages of feedgrounds

Several consequences of the elk feeding programs in the Jackson herd unit may be viewed as advantages by various groups of people. These include (1) attraction for tourists, (2) potential to increase herd size above that constrained by available winter range, (3) improved recruitment and reduced mortality and consequently higher sustainable yields for hunters, (4) reduced competition for other wildlife because elk are concentrated at feedgrounds, (5) reduced damage to private property, for example, haystacks and trees, and (6) improved ability to handle elk for research and disease control.

Winter concentrations of elk on the National Elk Refuge provide one of the most glorious winter wildlife spectacles anywhere. To ride a hay sled amongst thousands of wintering elk while viewing the magnificent Teton Range as a backdrop is an experience that few could forget. Such opportunities for wildlife viewing improve public appreciation for wildlife and allow an unequaled forum for public education about elk. The National Elk Refuge unquestionably helps to attract tourists who come to Jackson Hole to ski and to enjoy other winter sports.

The prevailing sentiment among wildlife biologists is that winter feeding is undesirable and should be avoided if at all possible (Lyon & Ward, 1982: 459). This view is stated particularly well by Boyd (1978) who insists that winter feeding can *only* act to the detriment of an elk herd and its habitat! Further, he states that 'After 65 years of expensive winter feeding, the Jackson elk herd is considerably smaller than when feeding began.' It is not clear to me that Boyd's statement is true, and certainly he does not have his facts right for the Jackson elk herd. Indeed, the feeding program has been expensive, but it is not true that the elk herd is considerably smaller now than when feeding began (see Figure 3.1). There is actually a statistically significant, although slight, positive trend in the number of elk wintering at the National Elk Refuge between 1912 and 1985 ($r_s = 0.272$, $n = 57$, $P < 0.025$). Maximum counts of elk actually fed between 1927 and 1985 does not show a significant temporal trend ($r_s = -0.091$, $n = 45$, $P > 0.2$).

Boyd (1978: 28) further claims that 'Without exception, reproduction in artificially fed elk herds is lower than in naturally subsisting herds.' This is simply not true. Recruitment for the artificially fed Jackson elk herd compares very favorably with recruitment from naturally subsisting herds throughout the Rocky Mountain West (Houston, 1982). Indeed, by maintaining weights with adequate supplemental feed, survival of calves may be enhanced (Thorne, Dean & Hepworth, 1976). This is supported by the fact that total hay fed on the National Elk Refuge and calf/cow ratios during the following year are positively correlated ($r = 0.766$, $n = 7$, $P < 0.05$) between 1936 and 1959 when feeding levels varied more than they have in recent years. Similarly, fecundity of white-tailed deer in Northern Michigan has been elevated by supplemental feeding (Ozoga, 1987).

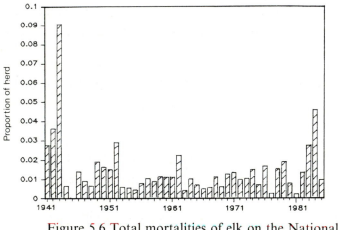

Figure 5.6 Total mortalities of elk on the National Elk Refuge, 1941–85.

One of the initial objectives for winter feeding of elk at the National Elk Refuge was to reduce overwinter mortality (Preble, 1911; Sheldon, 1927). Refuge feeding operations are remarkably successful in this regard, as mortality on the Refuge is very low. Since 1944, average mortality has been only 1.2% (Figure 5.6), which is slightly lower than the 1.8% adult mortality suffered by cattle on ranches in Wyoming between 1980 and 1984 (Habib, 1984). This is remarkable when one considers that elk mortalities include deaths to disease, old age, injuries sustained in fighting, cripples from hunting season, etc. Cattle likely to die of old age are often culled by ranchers, some diseases can be treated before becoming fatal, and deaths to fighting and wounds imposed by hunters are certainly less common in domestic livestock than in elk.

Ozoga & Verme (1982) state that: 'Traditional principles aside, perhaps biologists should stop viewing artificial feeding as professional heresy and realize that when properly administered it can serve as a valuable tool in the current state of the art . . . When properly conducted, supplemental feeding provides a feasible method of maintaining a reasonably large deer herd in good physical condition with minimal damage to the range.' Yet, Ozoga & Verme (1982) emphasize that for winter feeding to be successful, adequate harvest must be maintained to check population size.

Disadvantages of feedgrounds

Artificial feeding creates the 'disadvantage' of encouraging 'further growth of the population and creating ever-increasing demands upon both artificial and natural food' (Robinson & Bolen, 1984). This can, of course, be circumvented with adequate harvesting (Ozoga & Verme, 1982). Demand for elk hunting and manipulation of harvests by the Wyoming Game and Fish Department has resulted in an adequate harvest from the Jackson elk herd, as evidenced by only a slight trend in numbers of elk on the National Elk Refuge since 1912.

But there are additional disadvantages to feeding for both elk and man. For elk, there is increased risk of disease transmission created by large concentrations of animals on winter feedgrounds. And there can be substantial impacts by elk on the habitats surrounding these feedgrounds that would probably not occur if the elk were free-ranging (Fretwell & Lucas, 1970).

Most subtle, however, are the consequences of feedgrounds for the resource manager. Concentrations of elk on feedgrounds offer considerable opportunity for public viewing and enjoyment. But with this comes a change in attitudes towards the elk. Elk appear to some to be like so many cattle pastured for winter. And too many people know how to manage cattle!

Parasites and diseases

One of the potential risks of concentrating large numbers of animals in a small area is the increased risk of disease transmission. Indeed, several diseases occur in the Jackson elk herd, and several of these may be particularly problematic in Jackson Hole because of high rates of transmission at feedgrounds (Murie, 1951b). Thorne *et al.* (1982) review diseases and parasites of elk in Wyoming, and Kistner *et al.* (1982) review parasites and diseases of elk in general. Here I only discuss four of possible management significance: brucellosis, cattle lungworm, necrotic stomatitis, and scabies.

Brucellosis

Of the many diseases and parasites occurring in the Jackson elk herd, clearly Bang's disease or brucellosis is of greatest management significance. The disease is usually caused by the bacterium *Brucella abortus*, although other species of *Brucella* cause identical pathology. The disease is highly contagious through milk, fetal membranes and uterine secretions, and is usually transmitted by licking an aborted fetus or dead calf, or through ingestion of contaminated vegetation (Thorne *et al.*, 1978). It infects a variety of species, particularly domestic cattle, swine and horses, elk, caribou (*Rangifer tarandus*) and bison. In addition, it can cause very serious disease in man, usually known as undulant fever.

In elk, brucellosis has very significant consequences for management because approximately 50% of the time it causes abortion or still birth of the first calf following infection (Thorne *et al.*, 1978). However, infected cows successfully raise later calves (Thorne, Morton & Ray, 1979). This results in reduced herd productivity, which may be viewed as detrimental for maximum-sustained-yield elk harvest objectives.

More serious, however, is the fact that brucellosis can be transmitted between elk and cattle (Thorne, Morton & Ray, 1979), although elk are more resistant to brucellosis than are cattle. Brucellosis causes abortion and infertility in cattle. Because of health concerns and economic impacts of the disease, the US Department of Agriculture maintains a vigorous brucellosis eradication program, administered by the Animal and Plant Health Inspection Service (APHIS). Law requires that infected cattle herds be depopulated, i.e., slaughtered, which may incur considerable expense and financial loss to the owners.

Contrary to remarks by Robbins, Redfearn & Stone (1982), refined serological techniques exist for reliable detection of brucellosis in elk. Identification of brucellosis infection in elk requires that two of four tests be positive; these include plate agglutination, complement fixation, rivanol, and *Brucella* buffered antigen rapid card tests (Thorne & Morton, 1975; Thorne, Morton & Ray, 1979). Morton, Thorne & Thomas (1981) evaluated the efficacy of each technique in detecting known brucellosis infections in elk. Complement fixation tests correctly identified 93% of elk that were later necropsied and shown to be infected by culturing *Brucella abortus*. Although the complement fixation test is most reliable, it is also the most technically difficult of the four tests to perform.

Brucellosis was first detected among bison from Yellowstone National Park in 1917 (Mohler, 1917), and among elk on the National Elk Refuge in 1930 (Murie, 1951c). It has since been found to be prevalent among elk on feedgrounds in northwestern Wyoming (Thorne, Morton & Ray, 1979). Table 5.4 lists the incidence of brucellosis among elk tested 1970–85. Frequency is highest among adult females, 40% of which test positively for the disease. However, only a few mature bulls have been tested. Yearling females show significantly ($P < 0.01$) lower incidence than older females, with 27% of 51 from the National Elk Refuge testing positive (Thorne, Morton & Thomas, 1978). Since most cows in a herd such as the Refuge wintering population probably become infected and one half lose a calf, brucellosis causes an estimated reduction in herd productivity of approximately 7% (Oldemeyer, Robbins & Smith, unpublished) to 12% (Thorne, Morton & Ray, 1979).

Because brucellosis is perpetuated in feedground and free-ranging elk in Wyoming, APHIS' nationwide program for the eradication of brucellosis cannot be accomplished until brucellosis in elk and other wild reservoirs is eliminated. Kistner *et al.* (1982) review three programs for eradication of brucellosis among bison and elk in northwestern Wyoming:

(1) Killing all bison and elk in the region, and then waiting long enough to restock until the bacteria had died. A similar drastic approach was used successfully to control foot and mouth disease in California when 22 000 deer were killed in Stanislaus National Forest during the 1920s (Hibler, 1981).

(2) Capture young bison and elk to be reared in enclosures, while destroying all other bison and elk. After 'cleaning the range', the penned animals could be tested and released if free from disease. The rationale for this program is that most individuals do not become chronically infected until sexually mature (Thorne *et al.*, 1978) and also captive animals could be vaccinated.

(3) Vaccination of all elk and bison calves with *Brucella abortus* strain 19 vaccine in order to accomplish herd immunity. Such a program may be 'doomed to failure', because protection by vaccine is not absolute, and assuring vaccination of a majority of calves may be impossible (Kistner *et al.*, 1982).

I find each of these programs to be impractical and undesirable. Given that such drastic measures might be required to achieve the present objectives of the National Brucellosis Eradication Program, the program's objectives obviously need to be reformulated. Despite an active brucellosis eradication program since 1934, brucellosis still persists in livestock even in states without elk or wild bison to reinfect livestock (Kistner *et al.*, 1982). But, this is most probably a consequence of cattle shipments from areas where brucellosis occurs.

Furthermore, the epizootiology of brucellosis is incompletely understood

Table 5.4. Prevalence of brucellosis in elk tested on the National Elk Refuge during winters of 1970–71 through 1984–85 (E. T. Thorne, unpublished). Totals include some individuals for which sex and age were not recorded.

Year	Mature females		Mature males		Yearling females		Yearling males		Calves		Total	
	% pos	N	% pos	N	% pos	N	% pos	N	% pos	N	% pos	N
1970–71	33	82	50	8	56	9	0	12	0	60	20	171
1971–72	46	37	50	2	45	11	33	3	0	12	32	92
1972–73	53	70	50	4	29	7	0	2	0	11	43	94
1973–74	50	106		0	18	11	0	8	22	18	40	143
1974–75	52	82	100	1	0	9	20	10	0	6	40	108
1975–76	35	102	50	2	0	4	0	4	6	48	25	160
1976–77	15	33		0		0		0	0	7	23	22
1977–78	47	179		0	20	5	25	4	0	38	40	216
1978–79	36	113		0	14	14	25	4	1	78	22	209
1979–80	19	72	0	1	0	4	29	14	0	1	20	92
1981–82	25	114		0	7	14	0	4	5	42	18	176
1984–85	20	10		0	33	3	33	3	0	34	8	50
Totals	39	1000	50	18	20	91	15	68	3	355	28	1533

(Kistner *et al.*, 1982), as evidenced by isolation of *B. abortus* from coyotes (*Canis latrans*; Davis *et al.*, 1979), wolves (*Canis lupus*), red fox (*Vulpes vulpes*), and grizzly bears (*Ursus arctos*; Neiland, 1970; Dieterich, 1981). Although occurrences in carnivores may be due to contact with infected livestock or cervids, and carnivores may not normally host the disease (Neiland, 1970, 1975), this has not been demonstrated (Dieterich, 1981). Grizzly bears can harbor the disease for a 'prolonged period' after experimental inoculation (Dieterich, 1981: 56). Yet, harboring the disease is of no consequence epidemiologically unless individuals shed large enough numbers of *Brucella* cells to be infectious.

Even though achieving herd immunity may not be practical, management actions can be taken to significantly reduce the risk of brucellosis transmission to cattle. Because transmission of the disease from elk to cattle is most probable through contact with an aborted fetus or still-born calf, an obvious solution is to minimize risks of such contact. This entails keeping cattle and bison off elk feedgrounds in winter when they could come in contact with aborted fetuses. Likewise, grazing of cattle on elk calving grounds should be avoided during the calving period (Thorne & Morton, 1975).

For the Jackson elk herd, such contacts between cattle and elk are most likely at (1) Uhl Hill in Grand Teton National Park where cattle are grazed on elk calving grounds during the calving period (Boyce & Sauer, 1978; Sauer & Boyce, 1979a; Sauer, 1980a, b), and (2) in the upper Gros Ventre River valley where a similar situation occurs (Long *et al.*, 1980). One solution is to prohibit cattle grazing before 30 June to ensure that viable *Brucella abortus* organisms are not available for acquisition by cattle. This recommendation was offered by Thorne & Morton in 1975, but to date has not received response from management authorities. APHIS currently recommends that ranchers keep cattle away from elk before 10 June (D. Woody, pers. comm.), although elk calving certainly extends beyond this date (Morrison, Trainer & Wright, 1959; Cole, 1969).

Another approach which will reduce the probability of infection passing from elk to cattle is to reduce the infection rate among elk by inoculation of calves and adult females. Increased resistance to brucellosis afforded by immunization may also reduce abortion rates and thereby reduce chances of contact by cattle with a highly contagious fetus. E. T. Thorne (unpublished) has launched a research program, partly funded by APHIS, to explore the feasibility of inoculating elk by using biobullets containing strain 19 *Brucella* vaccine shot from an air gun. On feedgrounds where elk cannot be readily approached, Thorne proposes to corral-trap calves at creep feeders for hand injection of the vaccine.

Research to date has indicated that biobullets can be used successfully to inoculate elk on feedgrounds. Full doses (relative to those prescribed for cattle) of strain 19 vaccine have been found to cause abortion among some cow elk (approx. 27%), but a reduced dose still provides immunity without abortion (Thorne & Anderson, 1984). Although it is not currently possible to assess which calves have already been inoculated with biobullets, multiple inoculations at reduced doses are not harmful to the elk (E. T. Thorne, pers. comm.). Inoculations can be conducted by Wyoming Game and Fish personnel employed to feed the elk without major investment of time or resources. Feedgrounds on which pens must be constructed will require greater capital investment and a larger field crew to trap elk and administer vaccine by hand.

If successful, the inoculation program certainly can reduce the risk of brucellosis transmission to cattle. However, it seems unlikely that the technique can achieve adequate herd immunity to eradicate the disease (Kistner *et al.*, 1982). Anderson & May (1985) have studied models of herd immunity from vaccination. They find that the proportion of each cohort that needs to be vaccinated is lowest for youngest animals, but for feedground elk, the youngest that calves are concentrated so that they could be inoculated is 6–7 months of age when they arrive at the feedgrounds. This increases the probability that the calves have already obtained the disease, although the fraction infected is certainly lowest among calves (Table 5.4).

Although adequate details on the epidemiology of brucellosis are not available, work with other diseases indicates that a rather high fraction of the herd must be vaccinated to achieve immunity. Values for other diseases range from 84% to 96% (Anderson & May, 1985). During the first year of Thorne's study, approximately 70% of the cow and calf elk on the study feedgrounds were inoculated. Although 70% vaccination of elk on feedgrounds may not be adequate to achieve herd immunity, continued efforts are hoped to eventually yield nearly 95% vaccination frequency for the herd (Thorne, pers. comm.). It remains to be seen if this is possible, particularly because not all elk winter on feedgrounds every year (see Chapter 3).

Vaccination is presumed to convey life-long immunity for elk, although test elk have only been monitored for two years (Thorne *et al.*, 1978). Also, to achieve herd immunity, Anderson & May (1985) suggest that vaccination should continue for several decades, and that no reservoirs of the disease remain endemic in nearby areas. Since elk and bison in Yellowstone National Park are infected with brucellosis (Rush, 1932; Meagher, 1973), these populations would have to be included in any successful inoculation program. Nevertheless, herd immunity is not a stated objective for the program; rather, a goal is to reduce the probability of transmission from elk to cattle.

Is brucellosis a significant management problem for the Jackson elk herd? I argue that it is indeed significant because transmission of brucellosis from elk to cattle can cause considerable economic loss for cattlemen. Currently, incidence of brucellosis in the United States is only 0.7% among cattle herds although in 1976 this resulted in at least $20 million annual loss to the cattle industry (Becton, 1977). Two outbreaks of brucellosis in cattle have occurred in northwestern Wyoming since 1982, for which no bovine source could be identified by APHIS, and it was thought that these may be due to transmission of the disease from elk (D. Woody, pers. comm.). One outbreak occurred on the Black Butte Ranch near Cora, Wyoming, which hosts an elk feedground, and the other was in a dairy herd near the Grey's River elk feedground in Star Valley, Wyoming. Such conflict should be avoided if possible for no other reason than to avoid entertaining suggested control measures such as those employed on the Stanislaus National Forest deer herd; or those used for badgers (*Meles meles*) in England where tuberculosis is transmitted to cattle (Cheeseman *et al.*, 1985); or those proposed for northwestern Wyoming by Kistner *et al.* (1982), which I listed above.

Perhaps it is adequate simply to warn ranchers of the risk of brucellosis, and to inform them of practices to minimize risk of transmission. Cattle are usually immunized for brucellosis, further reducing the risk of transmission, although immunization does not offer absolute protection. To date brucellosis has not been a serious problem within the Jackson herd unit, and the economic benefits of using calving areas prior to the recommended date of 30 June may be worth the risk to an individual rancher. Of course, if brucellosis should be contracted, the consequences of APHIS regulations may impact all ranchers in Wyoming.

I submit that the detailed research on elk brucellosis in northwestern Wyoming is some of the most thorough and comprehensive disease research ever conducted on a large wild-mammal species, and that this work has clear management implications. I concur with Thorne (1984) that, in view of the facts, to dispute the occurrence and importance of brucellosis in elk in the Jackson herd 'would only seem to perpetuate problems and will never enhance the credibility of managers'.

Lungworm

Although very common amongst elk in Jackson Hole, the metastrongylid nematode, *Dictyocaulus viviparus* (= *D. hadweni*) appears to cause mortality in relatively few animals. Pathology from the disease can certainly be severe when infection is heavy (Kistner *et al.*, 1982) and lungworm infections may predispose animals to bacterial infection (Thorne, Kingston, Jolly & Bergstrom, 1982).

Incidence of lungworm varies substantially among different sex and age classes of elk on the National Elk Refuge, with 70% ($n = 50$) yearlings infected but only 16% ($n = 340$) infection among mature cows (Bergstrom, 1975; Bergstrom & Robbins, 1979). Incidence of infections also varies seasonally, with peak infections of 80% found in fecal samples from Grand Teton National Park in May. In lactating cows lungworm infections persist until rut, although it is usually not detectable in other age and sex groups by late summer (R. Bergstrom, unpublished).

Possible density dependence or at least elevational differences in lungworm infection is implied by low incidence of 13–16% in summer samples collected on Big Game Ridge and the immediate vicinity in the northern Teton Wilderness. This is in comparison with high density populations in the central valley of Grand Teton National Park, where 30–40% were positive for lungworm in summer (Bergstrom & Robbins, 1979). Recent observations by Drs R. Bergstrom and D. Worley on lungworm in elk from the Gibbon Geyser Basin in Yellowstone National Park are consistent with data on seasonal variation in elk from the National Elk Refuge that summer immediately south of Signal Mountain in Grand Teton National Park; thereby attendance at feedgrounds is not required to sustain lungworm infections.

Since strains of lungworm in cattle will not persist in elk, and strains from elk do not persist in cattle (Bergstrom & Worley, unpublished), transmission of lungworm from elk to cattle probably is not a management concern. At one time, *Dictyocaulus viviparus* from cattle was considered to be a different species from *D. hadweni* in elk (Chapin, 1925). Because the two species are morphologically indistinguishable, classification as different strains may be more appropriate.

Since severe debilitating or fatal pathology directly due to lungworm infections has not been observed at the National Elk Refuge (Bergstom & Robbins, 1979), little consideration has been given to controlling the parasite. Moderate infections may reduce stamina, but 'few elk die from lowered tidal or residual air capacity' (Thorne *et al.*, 1982). If it should be deemed desirable, treatment can be administered by adding anthelmintics to supplemental food.

Necrotic stomatitis

During the winter of 1943, large numbers of elk died on the National Elk Refuge (Figure 5.6). Most of these mortalities were attributed to necrotic stomatitis (*Fusobacterium necrophorum*), a bacterial disease that can be epidemic on winter ranges (Murie, 1951c). The disease is usually fatal, inflicting calves at higher frequencies than adults. The same bacterium causes foot rot, which was also very common on the National Elk Refuge during earlier years (Murie, 1951c).

Occurrence of necrotic stomatitis seemed to be correlated with hay diets containing foxtail barley (*Hordeum jubatum*), which contains sharp awns that penetrate the inside of the mouth and throat, predisposing animals to bacterial infection (Murie, 1951c). After 1943, only hay relatively free of foxtail barley has been fed, and as a consequence, necrotic stomatitis has only occasionally been observed (Robbins, Redfearn & Stone, 1982). Improved nutrition from alfalfa hay compared to grass hay since 1943 may also have helped reduce the incidence of the disease on the Refuge (Kistner *et al.*, 1982), as evidenced by declining incidence of foot rot (E. T. Thorne, pers. comm.). Elk fed inadequate rations during feeding trials were more likely to suffer foot rot (Thorne, unpublished).

Scabies

Two external parasites are particularly common on elk wintering on the National Elk Refuge: (1) winter ticks, *Dermacentor albipictus*, which are not known to cause mortalities (Murie, 1951c; Robbins, Redfearn & Stone, 1982), and (2) psoroptic mites, *Psoroptes* sp., which cause scabies. Heavy scabies infestations result in extensive hair loss (alopecia) which can increase cold stress in winter and ultimately lead to death (Figure 5.7). Transmission of the mites to other individuals is usually by direct contact, although they may be transmitted at bedding sites.

On the National Elk Refuge, scabies has caused a substantial fraction of the overwinter mortality some years. During the winter of 1982–83, 62% ($n = 101$) of the mature bulls dying on the Refuge during winter were infested with scabies (Figure 5.8; Smith, 1985). In addition, several other mature bulls that died may have had scabies, but condition of carcasses when found precluded determining if scabies were present. If these bulls had scabies, its incidence among bulls dying on the Refuge may have been as high as 95%. Other than for mature males, the incidence of scabies was low (ca. 3%) in other elk dying on the Refuge during winter. Only two of 29 cows and none of eight yearlings or 29 calves had scabies.

The incidence of alopecia among live elk classified on the Refuge during the winter of 1982–83 was similar to that for dead elk; i.e., an exceptionally high incidence among mature bulls (Table 5.5). Elk with light alopecia do not usually possess mite infestations, and only severe alopecia can be considered indicative of scabies (B. Smith, unpublished).

Thermal stress associated with hair loss is suggested by the temporal

Figure 5.7 A scabied bull elk with severe alopecia adorned with a radio-transmitter collar, 14 March 1986. Of 15 scabied bulls radio-equipped during winter 1985–86, at least 10 summered in Grand Teton National Park during summer 1986.

distribution of scabies-related mortalities on the Refuge. Most of the infested mature bulls died during January, the coldest month of the year (Figure 5.9).

The incidence of scabies appears to be highest for elk which summer in Grand Teton National Park. Lean, mature bulls with high incidence of alopecia are amongst the first arrivals to the Refuge in autumn. Evidence that these are primarily elk that summer in the Park comes from (1) at least 10 of 15 scabied bulls radio-collared on the Refuge in spring 1986 summered in the Park (G. Roby, unpublished), (2) radio-telemetry studies showing that early arrivals onto the National Elk Refuge are usually Park elk (Bruce Smith, unpublished), (3) the timing of track counts, (4) aerial observations during fall migration, and (5) a correspondence between the high fraction of bulls amongst early arrivals onto the National Elk Refuge in autumn, and the high frequency of bulls among elk that summer in the Park. The first 1350 elk to appear on the Refuge during November 1982 consisted of 48% bulls, which was remarkably close to the proportion of bulls in the Park (47.2%, $n = 398$) during July and August (Figure 3.20).

Murie (1951c: 169) implies that scabies may reflect poor range condition and

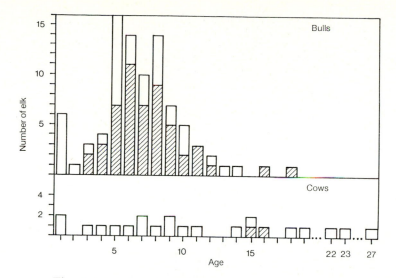

Figure 5.8 Age distribution of elk mortalities on the National Elk Refuge, 1982–83, showing the distribution of individuals known to have scabies (adapted from Smith, 1985). Cross-hatched areas are individuals known to have scabies. Open ares include 25 bulls for which insufficient hide remained to make a determination.

that animals in good physical condition are unlikely to be infested. However, there is no indication of deteriorating range condition or forage availability in the Park. Indeed, recruitment rates have increased since 1975 in GTNP (see Chapter 3), suggesting improved nutrition consequent to increased harvests in the Park and Refuge beginning in 1975 or 1977. Yet, recruitment rates may be independent of range condition in areas used by bulls, and summer distribution in Grand Teton National Park is different for mature bulls than for cows and calves (Martinka, 1969; see also Clutton-Brock, Guinness & Albon, 1982). As

Table 5.5. *Incidence of alopecia observed in the National Elk Refuge wintering elk herd, winter 1983–84 (from Smith, 1985).*

Class	Number	Number (%) with light alopecia	Number (%) with severe alopecia	Total number (%) with alopecia
Calf[a]	706	7 (1.0)	0	7 (1.0)
Cow[a]	2886	92 (3.2)	2 (0.1)	94 (3.3)
Spike[b]	345	33 (9.6)	0	33 (9.6)
Bull[a]	1050	240 (22.9)	52 (5.0)	292 (27.8)
Total	4987	372 (7.5)	54 (1.1)	426 (8.5)

[a] From February 6 count.
[b] From January 31 count.

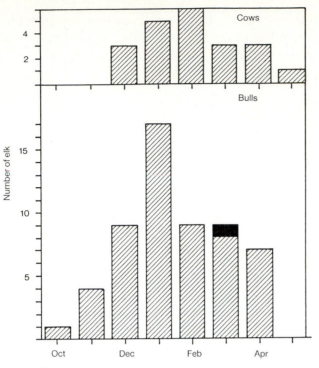

Figure 5.9 Temporal distribution of overwinter elk mortalities on the National Elk Refuge for the winter of 1982–83 (adapted from Smith, 1984). Solid area is a single spike bull.

suggested by Smith (1985), nutrition on winter range does not appear to have influenced scabies incidence since there is no significant correlation between the percentage of mature bulls among total mortalities (1975–83) and the number of days of supplemental feeding on the National Elk Refuge ($r = 0.364$, $n = 7$, $P > 0.1$), or the daily ration of feed provided ($r = -0.303$, $n = 7$, $P > 0.1$). Also, most mature bulls dying on the Refuge had massive antlers (Smith, 1985), which are unlikely to develop if nutrition is poor (Bubenik, 1982), at least during April to July when growth of antlers occurs (see Severinghaus & Moen, 1983).

The high scabies-related mortality among bull elk is more likely a consequence of malnourishment and physiological stress during the rut rather than poor nutrition during summer or winter (Smith, 1985). The highest frequency of scabies-related mortality occurs among prime-age bulls that would be most heavily involved in the rut (Figure 5.8). As shown in Figure 3.20, there is an exceptionally high proportion of mature bulls in GTNP resulting from the herd reduction program which focused primarily on antlerless elk. The intensity and stress of rutting may be higher in the Park where greater numbers of bulls compete for limited numbers of cows. This interpretation is reinforced by the fact that the percentage of bulls among Refuge winter mortalities increases with a slope greater than unity ($b = 2.19$) as a function of the percentage of bulls on the Refuge during winter (Figure 5.10; $P < 0.01$).

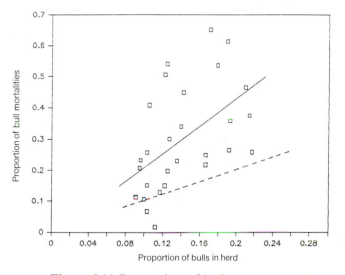

Figure 5.10 Proportion of bulls among overwinter mortalities on the NER as a function of the proportion of bulls among all elk classified on the Refuge during the same winter, 1952–86. Dashed line is the expected relationship if winter mortalities were a random sample of elk on the Refuge. Least squares regression: arcsine (sqrt (proportion bull mortalities)) = 1.87 (arcsine (sqrt (proportion bulls in herd)) − 0.153; $r = 0.545$, $P = 0.002$.

Impacts on habitats

When elk concentrate on feedgrounds, they often severely impact surrounding habitats. When not on the feedline, elk spend a high fraction of their time resting (Smith & Robbins, 1984). But when natural forage is nearby, elk may do substantial damage by debarking trees and heavily browsing trees and shrubs (Thorne & Butler, 1976). The most serious impacts are on aspen trees less than 4 m tall, but even conifers of low preference may be severely damaged by elk browsing (DeByle, 1985a). Pronounced browse lines are common in forests adjacent to feedgrounds. This alteration in the forest understory may substantially change avian faunas (Flack, 1976; Casey & Hein, 1983). In addition, this browsing reduces the value of forest habitats surrounding feedgrounds for thermal or hiding cover (Thomas, 1979).

Decadent aspen stands near feedgrounds (Figure 5.11) are unlikely to recover if not protected from elk browsing (Krebill, 1972; DeByle, 1985a).

Local damage can be severe, although some areas even near feedgrounds may escape elk browsing. Less than 2 km west of the Alkali Creek feedground aspen stands surrounded by open sagebrush steppe show vigorous growth of shoots and suckers.

Elk which are fed hay pellets are more likely to forage than elk fed baled hay because passage rates are higher for pellets and consequently within a few hours elk are hungry again (Thorne & Butler, 1976). This might be a particularly significant risk on some state-owned feedgrounds where debarking of trees and localized range deterioration is evident. To reduce this problem, Thorne & Butler recommended that feeding occur twice daily.

Advisability of this recommendation for the National Elk Refuge is not clear. Smith & Robbins (1984: Tables 5 and 6) found that fecal samples from elk fed baled alfalfa showed greater consumption of natural forage than those fed alfalfa pellets. Although free-ranging is discouraged on many state feedgrounds, in part to keep elk out of private haystacks, free-ranging for natural forage is actually encouraged on the National Elk Refuge where 12 000–13 000 tons of standing herbaceous forage is produced annually. Smith & Robbins (1984: 63–4) argue that faster passage time for pelletted hay may be beneficial if it allows elk to spend more time foraging on the Refuge and

Figure 5.11 Alkali Creek feedground in the Gros Ventre River valley, showing a severely browsed aspen stand.

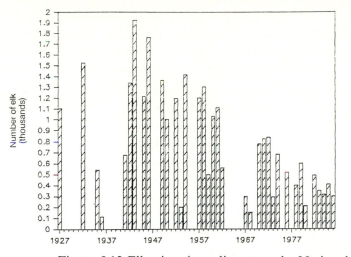

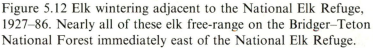

Figure 5.12 Elk wintering adjacent to the National Elk Refuge, 1927–86. Nearly all of these elk free-range on the Bridger–Teton National Forest immediately east of the National Elk Refuge.

consequently maintaining higher nutrition. Another drawback to feeding twice daily is that it will probably increase competition among elk on feedlines (Smith & Robbins, 1984) because smaller feedings allow feedlines to be consumed or dominated before subordinates have had an opportunity to feed.

Extensive foraging by elk is of concern because of possible competition with mule deer and bighorn sheep wintering on and adjacent to the National Elk Refuge. However, numbers of bighorn sheep and recruitment of lambs in the Curtis Canyon area have not declined subsequent to the introduction of pelleted hay at the National Elk Refuge. Furthermore, elk foraging adjacent to the Refuge have actually declined in recent years (Figure 5.12). Therefore, some of the concerns over the shift to feeding pelleted hay may be relaxed.

Would another feedground help?

The Jackson Hole Guides and Outfitters Association and the Concerned Citizens for the Elk Committee have recommended establishing an additional feedground along the Buffalo Fork, either at Burro Hill or further east. Since migratory Yellowstone–Teton Wilderness elk can potentially mix with GTNP elk during migration to the National Elk Refuge, separate harvests for these herd segments cannot be assured. Holding some of the Yellowstone–Teton Wilderness elk by feeding them at Burro Hill would allow greater control over harvest and feeding.

This has been done in the past. In 1970, the Wyoming Game and Fish Department fed elk at Burro Hill as an emergency measure to save approximately 250 elk that failed to migrate on to the Refuge. Some of the elk were trapped, tagged, and trucked to the National Elk Refuge, but the next winter the same elk were back at Burro Hill (I. J. Yorgason, unpublished). Because there was no interest in establishing a new permanent feedground at Burro Hill, permits were issued and the Burro Hill herd was culled (F. Petera, pers. comm.). Again in 1986, approximately 200 elk were provided emergency feed at Burro Hill by the Wyoming Game and Fish Department.

There are five serious objections to the proposed feedground location. First, Burro Hill is an elk calving ground (Altmann, 1952; Anderson, 1958). Concentrating a winter herd in the confines of the Buffalo Fork valley would result in localized habitat impacts like those at Alkali Creek (Figure 5.11) and would probably reduce hiding cover for calves, which is thought to be a key element of calving habitat (Skovlin, 1982).

Second, the willow flat in the Buffalo Fork valley is winter range for approximately 150 moose. Competition with elk likely would reduce the carrying capacity of this moose winter range. Elk will certainly browse heavily on willows (Hayden-Wing, 1979; Hobbs *et al.*, 1979; Marcum, 1979). There is no way to ensure that competition between elk and moose would not become extensive, particularly in mild winters when elk might free-range more than during heavy snow winters.

Third, soils in the Burro Hill area appear particularly sensitive to trampling disturbance. When elk were fed there in the past, trailing and erosion were very pronounced by spring (J. Yorgason, pers. comm.).

Fourth, cattle spend the winter on several ranches in the Buffalo Fork valley, consequently there is a possibility for transmitting brucellosis between elk and cattle. Cattle are most likely to contract brucellosis from elk during winter when elk calves are aborted (Thorne & Morton, 1975). Elk calves aborted on the National Elk Refuge have no chance of being contacted by cattle because cattle are not permitted on the Refuge.

Finally, grizzly bears use the Buffalo valley during early spring after emerging from their dens. Maintenance of a feedground in the Buffalo valley increases the chance for human/bear conflict. Such conflicts with grizzly bears should be avoided whenever possible.

Although I think that the suggestion of another feedground to separate Park elk from 'migratory' segments is an interesting one, I find objections to the Buffalo Fork site. Operations at the National Elk Refuge are well designed to handle large numbers of wintering elk, and no obvious alternative sites seem appropriate.

Is supplemental feeding necessary?

Wilbrecht & Robbins (1979: 252) suggested that 'If all the lands now under Refuge administration plus the adjacent National Forest lands, had been available in 1912, the program of winter feeding of elk might never have evolved. Now that these lands are available, it is feasible to manage the Refuge as winter range.' Although Wilbrecht & Robbins were posing this as a rhetorical statement in a vein of speculation, many, if not most, wildlife biologists consider winter feeding to be undesirable under almost any circumstances. Some consider all the problems associated with management of the Jackson herd as a direct consequence of a feeding program that should have never been initiated in the first place (Boyd, 1978; Beetle, 1979).

To understand these attitudes, the underlying premises must be considered. It is generally accepted that overwinter mortality often limits ungulate populations (Peek, 1986). When animals are dying during winter, the natural response is to do something about it, and a frequent reaction is to provide supplemental feed. Wildlife managers have realized for some time that enhancing the survival of natural populations typically leads to other problems. Either animals die of something else because other factors limit the population, or if winter mortality is truly limiting the population, winter feeding must be sustained year after year to avoid even more extensive losses later on.

Therefore, wildlife biologists generally believe that winter feeding is an inappropriate reaction to ungulates starving during winter and should be avoided. Some overwinter mortality is natural and healthy because it maintains natural selection on the herd. Winter feeding can only lead to even higher overwinter mortality in future years because winter feeding increases population levels and thereby increases the amount of natural or supplemental food needed during winter.

But does this rationale apply to the Jackson elk herd? Some of it does, but there are certainly extenuating circumstances. In one sense, it seems ironic that the same elk that summer in one of North America's largest and most spectacular wilderness areas migrate out in winter to be fed on feedlots like so many cattle! But at the same time, the rationale for supplemental feeding at the National Elk Refuge seems logical. That is, since man reduced the natural carrying capacity for the Jackson elk herd by developing parts of the winter range, if the original large elk herd is to be maintained for people to enjoy, that winter range must be replaced or supplemental feed must be substituted during winter. For the most part, other seasonal ranges remain intact; it is primarily winter range that man has impacted.

Given the fact that it is winter range that is being replaced by feeding elk in

Jackson Hole, I find Lovaas' (1970) perspective on winter feeding to be an overgeneralization: 'Feeding hay to elk is an attempt to have more elk than the natural range will support. There is no other possible reason to feed. But feeding can, at best, only compound the existing imbalance between elk and range. Feeding hay to elk is not successful.' Under some circumstances, I suspect that Lovaas' statement may be valid. Yet, I think that it is critical to recognize a distinction between a resident elk population and a migratory one with seasonal ranges like the Jackson elk herd. For the Jackson elk herd, only the winter range is out of 'balance', and this is due to human encroachment.

Increased size of the National Elk Refuge and improved forage production on Refuge lands has moved toward replacing winter range earlier usurped by man, although much of the Refuge still receives greater snow accumulations than on some of the developed lands in the valley, for example, south-facing slopes on East and West Gros Ventre Buttes. If the Refuge lands had been set aside in 1912 as Wilbrecht & Robbins (1979) considered, there may have never been the need to initiate the extensive and expensive winter feeding program that now exists. But if the feeding program had not been initiated in 1912, it is unlikely that there would be as many elk in Jackson Hole as there are now. Nor would there be the annual yield of elk for hunters. Nor would Jackson Hole have the appeal to visitors that it now maintains.

During the past 30 years, the Jackson elk herd has fluctuated less than before (Figure 3.1), due in part to a more consistent feeding program (Figure 5.2). In a sense, the ecosystem has lost some of its dynamic nature. Without the Refuge and supplemental feeding during winter, the size of the herd would fluctuate more widely. Such fluctuations would generate management problems; for example, periodic heavy mortality would have occurred that would have been very apparent to the public, and elk would have come into greater conflict with ranching operations and with residential and other developed areas. All these things would probably still occur without artificial feeding unless the herd was reduced below current objectives.

Some authors imply that the elk herd would be better off without the winter feeding program (Boyd, 1978). At the present time, because of strong public sentiment favoring feedgrounds, termination of supplemental feeding programs for the Jackson elk herd is not realistic. 'Sociopolitical considerations rather than principles of balanced resource management dictate the need' for winter feeding on the National Elk Refuge (Robbins, Redfearn & Stone, 1982: 500).

Despite these views, I maintain that the winter feeding program is justified on 'principles of balanced resource management'. Therefore, I see no reason to consider termination or even a scaling down of the winter feeding program. It is expensive, costing in excess of $250 000 annually for feed alone. But these costs are offset by (1) benefits to tourism in Jackson Hole, (2) increased harvest of elk by hunters, and (3) enhanced educational and research opportunity.

Summary

1. Elk have been fed hay on the National Elk Refuge in all but nine mild winters since 1912. Winter feeding for the Jackson elk herd is justified because natural winter range has been usurped by cattle ranching and by the town of Jackson.

2. Winter feeding on the National Elk Refuge typically begins in January and extends through late March or early April. On state feedgrounds, feeding usually extends later in the spring, often through April.

3. Local Boy Scouts assist in the collection and sale of elk antlers from the National Elk Refuge. Proceeds from the sale of antlers is used to assist with the purchase of winter feed for elk.

4. Feeding trials indicate that daily rations of 3.4 kg of pelleted alfalfa hay will maintain body weights of elk in winter. Lower rations may be acceptable because weight maintenance may not be necessary to satisfy management objectives for the feeding program.

5. Brucellosis occurs in approximately 30% of the elk on the National Elk Refuge, and may be partly a consequence of the winter feeding program because animals are concentrated and the probability of disease transmission is high. This is a serious management concern because of the risk of transmission of brucellosis to cattle. Minimizing contact between elk and cattle during late winter and spring will reduce the probability that brucellosis will be transmitted to cattle. Cattle should be kept off elk calving areas until late June.

Chapter 6.

Managing habitats on the National Forest

Despite the supplemental feeding program, 'over the long term, habitat is the key to determining the success or failure of an elk herd. All other factors are, in the final analysis, of much less significance' (Thomas & Sirmon, 1985). Elk hunters and others interested in the Jackson elk herd seem to overlook the significance of elk habitats. Other issues seem more immediate, for example, hunting regulations, winter feeding, Park herd reduction programs. Similarly, much of the management for the Jackson elk herd over the past 35 years superficially appears to have been by the Wyoming Game and Fish Department, the National Elk Refuge, and Grand Teton National Park. However, I submit that decisions by the Forest Service have had enormous effects on elk habitats, and that in the future, the greatest management decisions for the Jackson elk herd will be on habitat management by Bridger–Teton National Forest.

In this chapter I review management by the US Forest Service on portions of Bridger–Teton National Forest which fall within the Jackson elk herd unit. This discussion mostly focuses on the three consumptive land uses which have greatest potential for being in conflict with management for elk habitats: (1) livestock grazing, which in this part of Wyoming is nearly entirely cattle grazing, (2) logging, and (3) oil and gas exploration and mining. Although the National Forest includes wildlife in its management plans, little habitat management specifically for wildlife has actually occurred. All too often, wildlife habitat management on Forest lands actually entails justification for logging or range improvements that also enhance livestock grazing.

Management by the Forest Service is extremely important to the future of the Jackson elk herd because most of the land occupied by the elk herd is within the Bridger–Teton National Forest. This encompasses a total area of 3985 km² within the Jackson herd unit, including lands within a portion of the Teton Wilderness (1680 km²), and roughly one half of the Gros Ventre Wilderness (approx. 580 km²).

Land uses on the Forest

The Forest Service manages its lands on a principle of 'multiple use', which attempts to balance numerous demands on natural resources for the greatest total benefit (Berntsen *et al.*, 1983). Under multiple use, controlled public use of the land is permitted for grazing, timbering, hunting, recreation and mining. Demands and priorities for resource management vary, and consequently, management prescriptions may differ considerably on various Forest Service lands. Because multiple use offers flexibility in management options for the Forest, there are great opportunities for elk management; but this flexibility can pose threats to elk habitats as well.

Current management of Bridger–Teton National Forest is ruled by a host of plans, including Unit Plans, Multiple Use Plans, and most recently a ten-year Timber Management Plan approved in 1979. These plans are to be superseded by the Forest's Land and Resource Management Plan (Bridger–Teton National Forest, 1986b), which is required to meet the National Forest Management Act of 1976. This plan has recently undergone public review, and may undergo considerable change before the final plan is approved.

The 1986 Forest Plan is a large and complex document. Areas of the Forest within the Jackson herd unit are divided into 65 separate areas with management proposed under 13 different prescriptions, depending upon resources and priorities for management in each area. These include areas in which many commodity uses will be allowed including mining, logging, and livestock grazing, to areas managed for wilderness values or for preservation of grizzly bears. Some areas are recognized as having specific semiprimitive and primitive recreation opportunities and are planned to be managed to maintain these qualities, and will remain mostly roadless.

Areas designated as wilderness within the Forest are open to hunting but are closed to vehicular access, timber harvest, and to some degree, mining. Portions of two large wilderness areas are included in the Jackson elk herd unit: the Teton Wilderness Area (1680 km² within the herd unit) which was established in the original Wilderness Act of 1964 (Public Law 88–577), and the Gros Ventre Wilderness Area (ca. 580 km² within Jackson herd unit) which was designated in 1984. Combined, these areas constitute 57% of the National Forest within the Jackson herd unit. Although existing oil, gas and mineral leases or claims could be mined within these Wilderness areas, no future leases or claims may occur, and elk habitats in these areas are reasonably secure from development threats. A let-burn policy has been established for naturally occurring fires within extensive areas of Teton Wilderness (Reese *et al.*, 1975).

Livestock grazing

Grazing of domestic livestock on Bridger–Teton National Forest is managed on numerous range allotments. Within each allotment, the Forest Service may issue grazing permits to ranchers who meet certain requirements (ownership of base ranch property and livestock). Most livestock grazing permits in the Jackson elk herd unit are in the Gros Ventre River and Spread Creek drainages, although three allotments totalling 148.6 km², occur within the Teton Wilderness Area (Buffalo Ranger District, 1973).

In 1919, the Forest Service recognized the crucial nature of elk winter range on the Forest east of the National Elk Refuge and on the upper Gros Ventre valley by a Supervisor's order which designated 546.5 km² of elk winter range (Wilbert, 1959). Trailing of 4500 cattle each year through the Gros Ventre valley was allowed to move cattle to and from higher elevation summer range, but otherwise the designated elk winter range was to be off-limits to cattle grazing.

The trailing plan was not strictly followed by ranchers, however, and trespass cattle used the elk winter range during most summers (Smith & Wilbert, 1959). Overuse of the range by elk and cattle led Craighead (1952) to conclude that no usable browse within the Jackson elk herd winter range could be classified in 'good condition'. By 1957 cattle grazing permits were being issued by the Forest Service within the original area set aside as elk winter range (Smith & Wilbert, 1959). Studies by Wilbert (1959), Smith (1961), and Yorgason (1963, 1969) demonstrated excessive use of both browse and herbaceous vegetation on the winter range by elk and cattle.

Formally, the Supervisor's Order of 1919 establishing the Gros Ventre elk winter range was not binding, and subsequent Forest Supervisors did not need to adhere to this Order. In 1973, the Forest Service proposed a management plan for the designated elk winter range which entailed a rest-rotation system where most areas in the grazing allotment would be grazed only every other year. Because trespass cattle grazing seemed unmanageable, the Wyoming Game and Fish Department consented to this grazing plan, hoping that controlled cattle use was preferable to the prevailing situation (J. Yorgason, pers. comm.). This agreement was later formalized by the Fish Creek Cattle and Horse Allotment Grazing Management Plan and associated Environmental Analysis Report which is on file in the Jackson Ranger District office of the Bridger–Teton National Forest.

To further clarify the extent of grazing to be permitted on elk winter range, on 11 December 1979, a formal Memorandum of Understanding was signed by the Bridger–Teton National Forest and the Wyoming Game and Fish Department. This document states that livestock use of the Breakneck and Haystack Units of the Fish Creek Allotment 'shall only be made to the extent that no detrimental effect is made to the elk winter range habitat requirements, and that the integrity of the basic intent of the 1919 range designation of elk winter range be maintained. If unresolvable conflicts between big game and domestic livestock develop, removal of domestic livestock from these critical big game ranges will be required by the Bridger–Teton National Forest.' In addition, it was agreed that permanent long-term trend studies would be established in 1980 to be re-read at least every five years to monitor effects of cattle and elk use on the range.

Since this agreement was signed, violations of grazing plans have occurred almost every year. Acccording to reports on file at the Ranger District office, salt has been placed on the Patrol Cabin and Alkali feedgrounds by livestock grazing permittees to attract and hold cattle, and gates closed by the Forest Service have been by-passed. Very heavy grazing by cattle has been documented in the Breakneck Unit, particularly on State lands surrounding the Patrol Cabin feedground (Jackson District Office allotment file, US Forest Service).

In correspondence to Wildlife Biologist Garvice Roby dated 10 November 1982, Joseph Kinsella, Forest Service District Ranger, acknowledged that 'Everyone agrees that over-grazing is occurring most years particularly in the riparian bottom of the North Fork of Fish Creek . . . along Red Creek, Beauty Park Creek and Lower Squaw Creek.' The North Fork of Fish Creek is heavily grazed by cattle every year, under a deferred rest-rotation system which allows for use every other year. Seasonally, these riparian zones offer some of the highest quality habitats for elk (Yorgason, 1963; Thomas, 1979: 41–4).

Persistently there is a lack of proper distribution and movement of allotted cattle to agreed upon rest-rotation areas within the grazing allotment. However, despite the strong wording of the 1979 Memorandum of Understanding, the Forest Service continues to allow grazing in the Breakneck Unit and there continue to be grazing problems on the Unit.

At the suggestion of the Forest Service, permittees have used salt for several years to attract cattle out of riparian areas onto adjacent ridges. Although this may reduce use of riparian areas, it increases cattle use on ridge habitats which receive the most extensive use by elk during winter and early spring (Wilbert, 1959; Smith, 1961; Yorgason, 1963). Wilbert (1959) and Yorgason (1969) showed that overwinter use of herbaceous vegetation by elk on ridges above the

Gros Ventre valley exceeded 90%. Because these ridges may blow free of snow or first become free of snow in spring, they offer particularly important habitat for elk.

Trend transects were not established in 1980 as agreed upon in the 1979 Memorandum of Understanding. However, two were established in 1974, three in 1981, one in 1982 and one in 1985. The two transects established on the Breakneck Burn in 1974 were read in July 1986, and these indicated improved range condition. Five other transects were read in July 1986 as well; these were not permanent transects, but rather were transects in the estimated vicinity of an old range analysis begun in 1959 and finished in 1961 (see Wilbert, 1959; Smith 1961). These transects also suggested improved range condition. Probable reasons for apparent improvement of range condition on the Gros Ventre elk winter range include (1) improved forage production due to over 2000 ha of burning in the area during the past 15 years, (2) decreased elk numbers on the range (Figure 3.2), and (3) small sample size in a good year for range production, for example, July 1986 rainfall at Jackson totalled 39 mm, which is 70% above normal. A true assessment of trends in range condition on the Gros Ventre will require considerably greater sampling effort in more habitats than just sagebrush habitats of less than 30% slope. In particular, riparian areas and ridgetops should also be sampled.

The Patrol Cabin feedground appears to be a persistent problem area, largely because it is state land and a central point in the elk winter range. Because of this focal point, the Jackson District Ranger office of the Forest Service and the permittees have encouraged the Wyoming Game and Fish Department to fence the property. By Wyoming state law, to exclude cattle, the Wyoming Game and Fish Department must fence its property. However, concern by the Wyoming Game and Fish Department is not solely for the feedground property, but rather for the crucial elk winter range in the entire area. Also, a fence is not desirable because it may influence elk movements.

In view of persistent grazing problems on the Gros Ventre elk winter range, I am troubled by management prescriptions for the Breakneck Unit in the draft ten-year Plan (Bridger–Teton National Forest, 1986b). Northern portions of the elk winter range have been designated for prescriptions 2B and 2C in the Plan on which 'permitted livestock may increase numbers'. As justification for increased livestock use on the National Forest, the Forest Service offers a claim that 'most of the elk are fed at State feedgrounds on or adjacent to the Forest' and therefore elk no longer require the range resources (Bridger–Teton National Forest, 1986b)! This is certainly unjustified, particularly on the Gros Ventre elk range.

Elk were being fed at the time of the Wilbert (1959) and Yorgason (1963,

1969) range studies, yet vegetation was almost completely consumed by spring in most areas of the winter range. Elk have been fed an average of 84 days each winter on the Gros Ventre feedgrounds since 1967 (Table 5.2), usually beginning in late January and terminating by 15 April. During mild winters of 1976–77 and 1980–81, elk free-ranged all winter and no supplemental feed was provided. Prior to feeding in early winter and after feeding in spring, elk make heavy use of native forage (Yorgason, 1963). Elk normally use riparian areas before supplemental feeding begins, and extensively use bluebunch wheatgrass on slopes both before and after feeding (Yorgason, pers. comm).

As reviewed in Chapter 4, elk experience more competition with domestic cattle than with any other large herbivore because of their similar diets and range preferences (Nelson, 1982: 418). Heavy cattle grazing of range resources on the Gros Ventre elk winter range is in competition with elk (Long *et al.*, 1980; Nelson, 1982; Mackie, 1985). It seems somewhat ironic that during January 1986, cattlemen in the Gros Ventre valley complained that the Wyoming Game and Fish Department was too late initiating supplemental feeding for elk at the Fish Creek feedgrounds, where feeding began 10 January; and as a consequence, elk invaded private haystacks consuming over 70 tons of hay (Associated Press, 1986a).

Jackson Hole has traditionally been dominated by ranching interests (Hocker & Clark, 1981; Righter, 1982). But strong public interest in the scenic and wildlife values of the valley have gradually taken priority over cattle grazing. The Draft Forest Plan acknowledges that agriculture is declining in the valley (Bridger–Teton National Forest, 1986b: II–43), yet the Plan proposes to increase grazing AUMs (animal unit months) on the Forest. To accomplish increased livestock use in the face of a declining ranching industry, the Plan proposes to 'encourage more ranchers to stay in ranching' (Bridger–Teton National Forest, 1986b: II–43). Certainly ranching is more compatible with scenic and wildlife values of the valley than commercial and residential development of these lands (Hocker & Clark, 1981).

Except for the fact that elk are regularly fed at state feedgrounds, the issue of grazing on the Gros Ventre is little different now than 45 years ago when Olaus Murie met with several area ranchers and guides to hear complaints about the perceived declining status of elk in the area. Murie's response was alarmingly apropos to the present situation:

> You people, I suppose, would want to establish permanent hay-feeding grounds up here to begin with. The soil is poor. We all know that natural forage is scarce. Cattle are driven through here every spring and fall, to and from the high summer ranges beyond. Elk are

expected to winter here, on poor range that is part of a cattle driveway, and you know what that means. Yet there is no other way to get those cattle through . . .

Look at those willows out there in the creek bottoms. They're pretty well used up. Now if you start feeding hay up here you will concentrate the animals more than ever. You will build up a bigger herd, until you have many more times the elk this area can support. We have found, too, that in addition to hay the animals will go after browse. So you will lose the willows and the forest reproduction. And in spite of hay you will still have winter losses. And you can have no moose up here then, no deer . . .

People do not want to provide enough natural range for wildlife. Sportsmen demand bigger and bigger game herds but do not trouble to provide living space for them in the way nature intended. They want to simply stuff the animals with hay, the easy way – and that is supposed to settle all problems. That's what's the trouble with the elk! (Murie & Murie, 1966: 176–7).

Murie's quote is from approximately 1942 (M. Murie, pers. comm.). After this time, elk were fed on the Gros Ventre feedgrounds only on an emergency basis until about 1956, when Anderson (1958: 161) reports that 86% of the elk on the winter range were provided supplemental feed. Although feeding has occurred nearly every year since 1960, at least half of the elk on the Gros Ventre winter range winter off feed (Straley, Roby & Johnson, 1983, 1984). The Gros Ventre River drainage affords the major wintering range for the Jackson elk herd (Figure 2.3), although since the 1950s it has been claimed that the winter and spring range is depleted (Wilson, 1958). Enforcing restrictions on livestock use of this crucial elk winter range will enhance the value of the area for elk. Other management practices such as burning, cutting and spraying, can also be used to increase forage production, as proposed in the Forest's draft ten-year Plan (Bridger–Teton National Forest, 1986b). Indeed, during the past 15 years, range condition may have improved in part due to over 2000 ha of controlled burning by the US Forest Service.

In Chapter 5, I argue that winter feeding of elk on the National Elk Refuge is justified because key winter range has been usurped by cattle ranching and residential developments. However, this justification cannot be applied to the Gros Ventre feedgrounds, and if there is any place in northwestern Wyoming where a reduction in winter feeding for elk could be justified, it is on the Gros Ventre winter range. Conflicts with cattle ranching in the valley will probably prevent this from happening (see Associated Press, 1986a).

Much of the 1986 Forest Plan is actually favorable to elk habitat. For example, under all alternative plans being considered by the Forest, several elk wintering areas are protected from cattle grazing. This includes much of the south Gros Ventre Winter Range, and the East Refuge, West Front and Horse Creek Big Game Winter Ranges. All evidence that I have reviewed indicates that livestock grazing with rest-rotation can be compatible with management for elk summer range on Bridger–Teton National Forest (see Chapter 4). No reduction in permits need be imposed on the Forest, and cattle use should be able to continue at current stocking levels.

However, enforcement of existing restrictions on livestock grazing is necessary and appropriate. *Controlled* livestock grazing can be compatible with management for elk populations on Bridger–Teton National Forest. The important point is that increased grazing levels on the Gros Ventre elk winter range should not be contemplated. Particularly, I contest the implication that winter range is no longer important for elk because elk are now provided supplemental winter feed at state feedgrounds (Bridger–Teton National Forest, 1986b: II–43).

Logging

Roughly 65% of Bridger–Teton National Forest is forested (Berntsen *et al.*, 1983). Lodgepole pine (*Pinus contorta*) predominates followed by subalpine fir (*Abies lasiocarpa*), Englemann spruce (*Picea engelmanii*), Douglas fir (*Pseudotsuga menziesii*) and aspen. Extensive logging in the Forest did not begin until the early- to mid-1960s when the US Forest Service escalated its efforts to secure timber harvests from National Forests. Since then, approximately 30 million board feet of timber has been harvested annually from Bridger–Teton National Forest, mostly through clear-cutting.

Several of the early timber harvests in the Forest were efforts to control infestations of mountain pine beetle which attacks lodgepole pine. Substantial portions of Spread Creek east of Grand Teton National Park were harvest-salvaged during the 1950s and early 1960s in an attempt to control epidemics of the mountain pine beetle.

Logging can be either beneficial or detrimental to elk habitat, depending upon the size, location and juxtaposition of the operation (Thomas, 1979). Extensive even-aged stands of lodgepole pine are poor elk habitat, and breaking up some of these stands can improve the vegetation for elk. For a few years after logging, clearcuts may offer excellent forage for elk, to the extent that elk can be considered a nuisance by silviculturalists attempting to secure regeneration on the clearcut (Lyon & Ward, 1982). However, no conflicts between elk and conifer recruitment have been reported on Bridger–Teton National Forest (D. Eggers, pers. comm.).

Bridger–Teton National Forest has employed a habitat effectiveness model

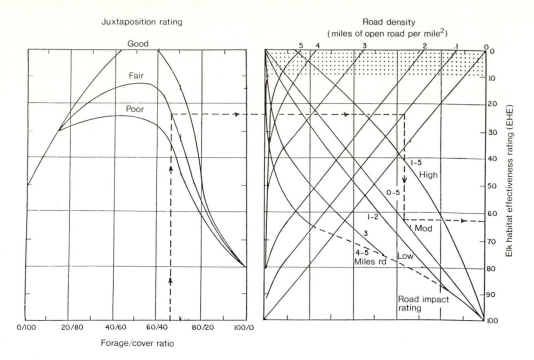

Figure 6.1 Elk habitat effectiveness (EHE) model used by Bridger–
Teton National Forest (Boss *et al.*, 1983; Leege, 1984). To evaluate the
EHE for an area, enter forage/cover ratio at bottom of left graph.
Follow upwards to appropriate juxtaposition curve. Then to the right
to desired road density curve, 0–5 miles of open road per mile². Finally,
bounce off the road impact curve to the right-hand axis to estimate an
EHE rating. When no roads exist, use only the moderate road impact
curve.

since 1979 for evaluating the value of areas for elk habitat (see Chapter 4). Basic
inputs into the model are (1) forage/security cover ratio, (2) juxtaposition of
forage and cover areas, and (3) roads (Leege, 1984). Based upon these three
factors, an assessment is made of the potential effectiveness of an area as elk
habitat (Figure 6.1). To illustrate the model, consider an area with a
forage/cover ratio of 65/35 with fair juxtaposition of forage and cover areas,
and approximately one mile of moderately used roads per mile². Beginning at
the bottom of the left graph in Figure 6.1, find 65/35, then follow up to the 'fair'
juxtaposition curve. Extend this point to the right until intercepting the one-
mile road density, then down to the moderate road use curve, and finally,
extend the line right to obtain an estimate of elk habitat effectiveness, here equal
to 62.

All timber sales proposed for the Jackson elk herd unit will require prior evaluation using this elk habitat effectiveness model. In its present form, however, the elk habitat effectiveness model has three major weaknesses. (1) The model does not include the consequences of livestock grazing. (2) The model is designed to evaluate summer habitats, and it does not include habitat requirements for seasonal migration. This is a particular concern for the Mt Leidy Highlands, which is scheduled for timber harvest and oil and gas exploration, but contains key migration routes to the National Elk Refuge and the Gros Ventre feedgrounds. Open 'forage' areas are relatively less important than security cover along migration routes. (3) There is a risk that the elk habitat effectiveness model could be used to justify timber harvest in roadless areas. By establishing a lower limit for habitat effectiveness for various segments of the Forest, timber harvest and road construction will eventually become dispersed throughout the Forest, at the expense of areas that presently have no roads or clearcuts.

Despite these weaknesses, as a first approximation the elk habitat effectiveness model appears to have great value for evaluating the consequences of logging operations on the value of elk habitats on the Forest. Most of the area within the Jackson elk herd has been assigned prescriptions which require high priority for wildlife habitat values, and I think that we can anticipate increased attention to the juxtaposition of timber harvests, roads, and size of clearcuts in a fashion that may enhance elk habitats.

The Forest Plan (Bridger–Teton National Forest, 1986a) explicitly prescribes that timber harvest units within the Forest will not exceed 400 m in width. Clearcuts will be interspersed with conifer forest cover patches of 10–25 ha in size and 400–600 m in width and length. Corridors of forest cover 200–400 m wide will be retained between patches of cover, and the distance along these corridors between cover patches will not exceed 400 m. Such restrictions on the size and interspersion of clearcuts will help to ensure that a high perimeter of forest edge occurs in the Forest, and will provide optimal habitat for elk.

Roads probably pose greater threats to elk management than does logging (Hieb, 1976). Increased road access into many areas of Bridger–Teton National Forest has reduced the value of these areas for elk habitat (Long *et al.*, 1980). Elk typically avoid roads, and road access increases elk hunter harvests to the extent that some local elk populations may be depleted, for example, the Togwotee area elk as discussed in Chapter 2.

Other guidelines for roads in elk habitats can help to minimize the impact of roads on elk. Minimizing straight stretches of road is helpful because curves in the road reduce visibility for elk and hunters. Locating roads adjacent to dense cover reduces the impact that roads will have on surrounding habitats. In open areas, road traffic may be visible to elk, and thereby avoided, for at least 2 km.

However, if the road passes through dense forest, the reduction in effective elk habitat may only extend for 150–200 m (Thomas, 1979).

The most effective way to manage roads in elk habitat is to close them. Road closures should be vigorously pursued in all situations where roads are in conflict with management objectives for elk. There is occasionally strong public resistance to road closures because recreational users of the forest are denied access. Much of this resistance can be circumvented if recreational access is never granted. Timber harvests can proceed without recreational access, and if the road is closed or destroyed immediately after timber harvest, recreational users will not learn to appreciate the opportunities offered by road access into an area.

Remarkably few road closures exist within the Jackson herd unit at the present time. Within the Buffalo District of the Forest there are 13 gates and about 10 gates closing roads in the Jackson District. Many of these gates are not to protect elk habitats, but rather to protect the road from use during periods when travel may damage the road, for example, when roads are wet in spring. Therefore, many of the gates are open seasonally and offer little enhancement of elk habitat. In other instances, road closures may be ineffective because gates may not adequately block access to the road.

The draft Forest Plan (Bridger–Teton National Forest, 1986b) proposes to be more aggressive in the future about closing and removing roads after timber sales. As an example of changing attitudes about roads on the Forest, the North Fork of Fish Creek timber sale for 5 million board feet of timber was sold in April 1986. To harvest this timber will require that the Forest construct 10.3 km of new road. However, after the timber harvesting is completed, 4.8 km of this road will be obliterated and made unusable. In addition, 9.7 km of other roads in the area will be destroyed with the remaining roads only open to snow machine traffic during winter when no elk are in the area.

In 1983, a team of resource managers evaluated resource management programs on Bridger–Teton National Forest (Berntsen *et al.*, 1983). The team noted that one of the most frequent complaints that was voiced was that 'it was difficult, if not impossible, to get information from the Forest Service about plans for future road locations'. The new ten-year Plan should alleviate this problem somewhat because the Plan explicitly lists 263.2 km of roads planned for construction with planned timber sales over the next ten years (Bridger–Teton National Forest, 1986b). The Forest cannot anticipate road construction for oil, gas and mineral exploration, but according to the Plan, these roads are to be destroyed after use.

Conflicts between elk habitats and logging are likely to decrease in the future. The preferred alternative for management of the Bridger–Teton National

Forest in its draft ten-year Plan calls for a reduction in timber sales on the forest by 20–30%. This is of great concern to the Louisiana Pacific Corporation that operates a 20-year-old saw mill in Dubois, Wyoming. Louisiana Pacific claims that it must be assured of 16 million board feet in timber sales within Bridger–Teton National Forest to keep its mill viable. Management under the new Plan cannot ensure this much timber for Louisiana Pacific. Furthermore, most of the scheduled harvests during the next ten years will be from southern segments of the Forest which are less economical because of the high costs to transport the logs to Dubois. It is expected that many of the timber sales in southern portions of the Forest will be bid to the Long Tree Timber Company in Afton, Wyoming. As a consequence of insufficient timber volume and possible competition with smaller logging firms, the large Louisiana-Pacific mill in Dubois may be closed (Prevedel, Robison & Iverson, 1985).

Closing the Louisiana-Pacific mill will have enormous impacts on the economy of Dubois since approximately 25% of its economy is based upon the Louisiana Pacific mill (Prevedel, Robison & Iverson, 1985). The Forest has encouraged the town to foster more tourism-based industry, and Shoshone National Forest has promised increased timber harvests for a few years to help phase the town out of its overdependence on a timber harvest economy. Although there is considerable empathy throughout the region for the people of Dubois who will be affected by the closure of the lumber mill, the Jackson Hole Area Chamber of Commerce polled its members in summer 1985 to find that a majority strongly favored management of the Forest for recreational, scenic and wildlife values rather than commodity uses (Heller, 1986).

Conservation groups question whether large-scale logging can be justified on northern portions of Bridger–Teton National Forest, because the area has particularly high value for recreation and wildlife. Economic justification for logging on the Forest does not appear to be sound. A substantial proportion of timber sales on the Forest are 'below-cost' sales, where it costs more to prepare a sale for harvesting than is returned to the Forest in revenues from sale of timber. Most of the costs resulting in below-cost sales are due to road construction into undeveloped areas. Preparation of below-cost sales means that the Forest Service is subsidizing the timber industry and associated employment base.

Recent timber sales have created considerable controversy. Two sales in particular are in important elk habitat in the Mt Leidy Highlands north of the Gros Ventre elk winter range: the Gunsight Pass sale and the Bull Creek–Harness Gulch sale. Environmental groups protested these sales in 1985 because of potential conflicts with wildlife habitat, and the Bridger–Teton Forest Supervisor postponed these sales until after release of the Forest Plan.

Louisiana-Pacific strongly protested the decreased timber sales on the Forest and as a consequence Regional Forester Stan Tixier in the Ogden, Utah, Regional Office ruled that Bridger–Teton must replace the previously offered sales with comparable quantities of timber from other sales within the Forest. A recently proposed timber sale in the Grouse Creek area of the Mt Leidy Highlands will also be controversial, because it is in an important migration route for elk migrating to winter feedgrounds.

I am optimistic about future logging on Bridger–Teton National Forest. The Forest Service is clearly making an effort to ensure that timber sales are compatible with elk habitat requirements, paying particular attention to forage/cover ratios and juxtaposition between forage-producing vegetation types and patches of trees which offer security cover. Also, the Forest has begun active programs to close or destroy logging roads after timber harvests have occurred. Logging performed in such a careful fashion can actually be beneficial for elk (Thomas, 1979).

My optimism is shaded by the current procedures for implementation of wildlife guidelines into management practice on the National Forest. For example, recommended timber sales in the draft Plan were prepared before review of the sales using the Forest's elk habitat effectiveness model, or similar input from the Forest's wildlife staff. This ultimately compromises the planning process and may result in lower priority for wildlife habitat. Effective Forest planning should incorporate habitat effectiveness whilst timber sales are initially being considered.

Oil and gas

Exploration for minerals, particularly oil and gas, on Bridger–Teton National Forest poses numerous threats to elk habitats and migration routes. Exploratory wells have been proposed in areas which are important elk habitats; for example, recently, Amoco Production Company has proposed to drill a well above Sohare Creek in the Mt Leidy Highlands (Thuermer, 1985). This area is in elk summer range, contains elk calving grounds and is along key migration routes to the Gros Ventre winter range. Therefore, the area is of high priority for elk management.

Although the draft Forest Plan protects crucial grizzly bear habitat and proposes a buffer to oil and gas exploration adjacent to Grand Teton National Park, most of the prescriptions in the Plan for the National Forest do not preclude mineral exploration and development. Exploration wells themselves need not be disruptive to elk habitat or migration routes, but displacement of elk along roads to well sites can result in substantial habitat losses. In the draft

Plan, roads to oil and gas exploration wells are scheduled for obliteration after exploration on most prescriptions in the herd unit (Bridger–Teton National Forest, 1986b).

The real threat, however, is the prospect that one of these wells might be successful, leading to full-scale oil field development. Such development in the Mt Leidy Highlands is incompatible with priority management for elk in the area. The Forest Plan offers little consolation for concerns over potential mineral development in Jackson Hole. Leases must be granted for exploratory drilling, although permission to drill presumably does not ensure that permission will be granted for mining. An Environmental Impact Assessment will be required before mining can proceed. It would be exceptional for an oil company to be denied permission for field development should an oil or gas strike be made and would almost certainly entail a court battle (Thuermer, 1986b). Local conservation groups feel that it is more appropriate to perform an environmental assessment dealing with potential field development prior to exploratory drilling. Such an assessment should deal with the potential biological and socio-economic consequences of development.

Phil Hocker, formerly a local environmentalist and Sierra Club officer, claims that oil and gas exploration offers the single greatest environmental threat to the Greater Yellowstone Ecosystem (Repanshek, 1986). Again, usually it is not the exploratory drilling that threatens elk habitats in Jackson Hole, but rather the risk of full-scale oil field development. According to Hocker, the Forest Service has not denied a single oil and gas exploration permit request in the Yellowstone ecosystem, and there have been 180 such requests. Thus far, over 40 wells have been drilled on the Forest, and all have been dry. Is it necessary or desirable to continue indefinitely to issue permits for oil and gas exploration in Jackson Hole?

Recreation

In the past, the Forest Service has not given high priority to wildlife or recreational uses of the Forest according to Tom Toman, Wyoming Game and Fish Regional Supervisor (Melnykovych, 1985). Rather, Toman claims that timber harvest has been the main priority for Forest Management. The draft ten-year Plan makes it clear that wildlife and recreational values will be given higher priority in the future, particularly within the Jackson herd unit. The Forest Service recognizes the importance of tourism and recreational use in Jackson Hole, and this is reflected in substantial reductions in planned timber sales in northern portions of the Forest.

Recreational use and elk habitat management may not be completely compatible, because recreationists may displace elk (Morgantini & Hudson, 1979; 1985). Two major areas of concern include road traffic and recreational use of winter ranges.

At a public meeting about proposed improvements to the Union Pass road, I was impressed by the strong public sentiment against road closures on the National Forest. Many people expressed the opinion that Forest lands belong to the public and that road access to these lands should be expected. Balancing public demands for recreational opportunity with optimal management for elk habitat will continue to create challenges for Forest managers. As discussed above, improved road management during and after logging or mineral exploration can minimize public objection and conflict with elk management.

In cooperation with the National Elk Refuge, the Forest Service has seasonally restricted recreational use of the Forest immediately east of the National Elk Refuge to minimize conflicts between wintering elk and skiers, snowmobilers and other recreationists. This ensures that disturbance of free-ranging elk adjacent to the Refuge is minimized.

For several years, however, conflicts occurred with cross-country skiers who used the Curtis Canyon trail east of the Refuge to access the Goodwin Lake ski hut near Jackson Peak in the Gros Ventre Wilderness Area. Elk and bighorn sheep using the Curtis Canyon area adjacent to the Refuge were frequently displaced by recreational users, particularly cross-country skiers. Resolution was secured in 1986 when an alternative access route to the ski hut was obtained along the south side of the Twin Creek Ranch. Ultimate fate of the Goodwin Lake ski cabin will be determined by the Gros Ventre Wilderness Management Plan which is expected for completion in 3–5 years.

Elk habitat management practices

Keys to elk habitat management on Bridger–Teton National Forest appear to be (1) proper road management, (2) timber harvest prescriptions, with primary focus on size and juxtaposition of clearcuts, (3) control over livestock grazing, and (4) fire management. The first three of these have been discussed relative to land uses.

Gruell (1979) strongly recommended that more fire management needed to take place on Bridger–Teton National Forest. Prescribed burning has been extensive on the Gros Ventre to control sagebrush (*Artemisia tridentata*), covering 2142 ha during the past 15 years. Although the intent for burns was mostly enhancement of livestock range, elk prefer these burned areas (Wilbert, 1959), and some burns cover slopes which are too steep for livestock, but receive elk use.

Burning offers potential for range improvement, but expectations of Gruell

(1979) have not been substantiated by experimental burns on Breakneck Ridge on the Gros Ventre elk winter range. Here, increase in aspen sucker production after burning was inadequate to escape heavy browsing by livestock and elk (Bartos & Mueggler, 1981; Hart, 1986). Much more extensive burning will be required to promote aspen reproduction in areas receiving heavy use by cattle and elk. Perhaps more importantly, however, fire can be a valuable tool for reducing encroachment by conifers into aspen stands (DeByle & Winokur, 1985).

There is often resistance among foresters to prescribed burning because of the fear that a prescribed burn will become a wildfire. Except for fires within specified areas of designated wilderness areas (Reese *et al.*, 1975), the draft ten-year Plan states that 'all wildfires will be controlled' (Bridger–Teton National Forest, 1986b: IV–59). But the Plan does propose to increase the acreage burned for wildlife habitat enhancement.

In addition to these 'key' management approaches, others merit consideration. For example, silvicultural practice may have important consequences for elk habitat. Berntsen *et al.* (1983) note that tree thinning was often undertaken when trees were 1.5–2 m tall and just beginning to provide excellent security cover for elk. Consequences of thinning to the effectiveness of cover can be alleviated in three ways. First, as in the cover/forage prescriptions, scheduling patches of young trees for thinning can ensure that large areas are not impacted. Second, allowing trees to become slightly taller before thinning, say 2.5–3 m, will reduce loss of security cover that occurs with thinning. Third, leaving trees at slightly higher density, for example, spacing trees 3 m apart rather than 4 m, will also maintain more security cover in thinned stands.

While tree thinning reduces security and thermal cover for elk, it usually enhances the production of forage. If a stand of trees occurs in an area with abundant cover for elk, cover considerations for thinning may be irrelevant. Priorities for thinning depend upon the juxtaposition of forage and cover types in the vicinity.

Opportunities for management of habitats are greater on the Forest than anywhere in Jackson Hole. This is true because of the flexibility in the Forest's management charge and also because Bridger–Teton National Forest hosts the greatest area of elk habitat within the Jackson herd unit, totalling about 3975 km^2 or 72% of the herd unit. 'Just how elk will fare in tomorrow's managed forests depends on how the forests are managed ... and how the people ... are managed' (Thomas & Sirmon, 1985).

Summary

1. Most of the land occupied by the Jackson elk herd is within Bridger–Teton National Forest, an area totalling 3975 km². Management of Forest lands is determined by a planning process. Bridger–Teton National Forest is in the midst of finalizing a new management plan.

2. Potential conflicts with commodity uses of Forest lands such as logging, mining and cattle grazing could be alleviated if wildlife was given higher priority and more direct input at the initial stages of the planning process.

3. Exploration for oil and gas does not present serious threats to elk habitats in Jackson Hole. However, it is the imminent threat of the development of oil and gas resources which presents the greatest negative threat to elk habitats in the Jackson herd unit, particularly in the Mt Leidy Highlands. Human disturbance associated with oil field development is likely to alter routes for elk migrating to the National Elk Refuge and winter feedgrounds on the Gros Ventre River.

4. Timber harvest has been extensive in some portions of the herd unit. Logging roads allow hunters easy access into important elk migration routes, resulting in local overharvests. Elk habitat effectiveness models have been used by the Forest Service to manage elk habitats by balancing open and forested habitats with road density. However, these models can also be used to justify logging in roadless areas. Existing elk habitat effectiveness models are not adequate for evaluating habitats along migration routes.

5. Negative consequences of logging and mining could be ameliorated by road closures.

6. Cattle grazing on the Gros Ventre winter range is poorly managed on the National Forest due to inadequate enforcement of cattle distribution prescriptions. As a consequence, there is reduced native forage for elk.

7. Approximately 2000 ha have been burned in the Gros Ventre valley to enhance cattle range. This has improved forage abundance and quality for elk as well.

Chapter 7.

Management by the Wyoming Game and Fish Department

Hunters prefer a large elk population to improve the probability of a successful hunt, and certainly tourists like to see elk during their visits to Jackson Hole. But not everyone would like to have more elk in Jackson Hole. Large numbers of elk can create problems. They are expensive to feed during winter, they frequently raid private haystacks consequently requiring that ranchers be compensated for damages, they compete for forage with domestic cattle, and they may damage aspen stands thereby reducing habitat values for some other wildlife species, for example, moose and mule deer. Determining a desirable size for the elk herd, therefore, requires that a number of interests be compromised.

Such compromises are largely determined by the Wyoming Game and Fish Department which controls culling in the herd unit. In addition, the Game and Fish Department plays a major role by feeding elk at the Gros Ventre feedgrounds, and by sharing the costs of winter feeding on the National Elk Refuge. In this chapter I review management of the Jackson elk herd by the Wyoming Game and Fish Department, with particular focus on hunting.

The Wyoming Game and Fish Department is charged with the management of all wildlife resources for the state of Wyoming. Its operations are overseen by a seven-member Commission appointed by the Governor. The primary tools used by the Department to manage the Jackson elk herd are (1) setting of hunting seasons and regulations for manipulating hunter harvest, (2) contributions to the elk feeding program at the National Elk Refuge, and (3) maintenance of three elk feedgrounds in the upper Gros Ventre River valley. In addition, the Department and US Forest Service biologists and managers collaborate in managing wildlife habitat on Bridger–Teton National Forest. The Department also cooperates with the National Park Service in formulating and carrying out the elk reduction program in portions of Grand Teton National Park.

Management by population objectives

The Wyoming Game and Fish Department currently facilitates decision making through formal planning. The Wyoming Game and Fish Department Planner coordinates wildlife planning for the state (Crowe, 1983). The Jackson elk herd, consisting of 13 hunt areas, is considered one unit for setting objectives on desired elk population size, recruitment rates, bull/cow ratios, number of hunters, hunter success rates, and total hunter recreation days. Hunting seasons and regulations are then adjusted to meet these objectives as closely as possible (Gasson, 1987).

Numerous considerations enter into the planning process and the setting of population objectives. These include (1) population dynamics and hunter success patterns for the herd over the past 5–10 years, (2) the area biologist's information and recommendations for the herd, (3) desires of the public, (4) estimates of sustainable yield for the population, (5) feeding costs, and (6) claims for damage by elk on private property (Crowe, 1983).

Existence of feedgrounds in the Jackson herd unit complicates the planning process over that for herds which are not fed, because feedground limits must be approved by the Wyoming Game and Fish Commission prior to obtaining public input. The local Department wildlife biologist recommends planning objectives to the District Supervisor after discussions with the US Forest Service, US National Park Service, and the US Fish and Wildlife Service. Input from the public, including a public meeting in Jackson, is also solicited prior to promulgating the management objectives (Crowe, 1983). Population objectives may be altered at any time, but are normally reevaluated every 3–5 years.

For the Jackson herd unit, the current harvest objective is 2200 elk (Figure 7.1) and the post-harvest population objective is 11 000 elk. This is to include no more than 7500 elk on the National Elk Refuge, 2182 at the three Gros Ventre feedgrounds, and an estimated 1300 elk not wintering on feedgrounds. The quota of 7500 elk for the National Elk Refuge was established long before the current planning process was initiated by the Wyoming Game and Fish Department, based upon a joint agreement by the US Fish and Wildlife Service and the Department. In 1979, the Jackson Hole Cooperative Elk Studies Group recommended that this objective be changed to an *average* of 7500 elk in the Refuge herd, but the Wyoming Game and Fish Commission decided to maintain 7500 as an *upper* limit.

Other objectives for the Jackson elk herd include post-season (i.e., after the hunting season) ratios of 40 calves/100 cows, and 25 bulls/100 cows (Straley, Roby & Johnson, 1983). Statewide objectives suggested in the most recent 'Strategic Plan' call for hunter success (i.e., the percentage of hunters that kill an

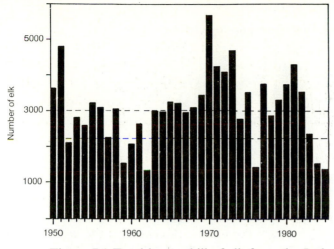

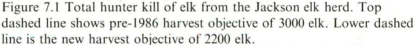

Figure 7.1 Total hunter kill of elk from the Jackson elk herd. Top dashed line shows pre-1986 harvest objective of 3000 elk. Lower dashed line is the new harvest objective of 2200 elk.

elk) at 24% and a hunter effort rate of 18.5 days of hunting for each elk harvested (Wyoming Game and Fish Department, 1983). Objectives set specifically for the Jackson herd area are currently 22% hunter success for 10 000 hunters, with 20 days for each elk killed totalling 44 000 hunter recreation days. Hunter success, number of hunters and total recreation days since 1971 are shown in Figures 7.2, 7.3 and 7.4.

Because most of the land in Jackson Hole is federal, the Department consults with other agencies before determining hunting seasons and regulations. In collaboration with the National Park Service, the Department sets seasons and license quotas for the herd reduction program in Grand Teton National Park. Likewise, input from the US Fish and Wildlife Service is obtained to assist in setting harvest regulations on the National Elk Refuge. Elk harvest on private lands is minor in Jackson Hole, perhaps of greatest significance on area 78 which sustains very little if any annual harvest.

To ensure public acceptance of elk management programs, public input is clearly very important in the planning process (Crowe, 1983). Yet, public sentiment may not always reflect sound management principles and can compromise population management objectives.

In 1984, a maximum of only 5055 elk wintered at the National Elk Refuge; the lowest count since 1960. Although this was in part due to a mild winter not forcing large numbers of elk onto the Refuge, every indication was that the population was lower than expected. Public reaction to the low numbers was strong. The general sentiment was that the herd should be allowed to increase.

(a)

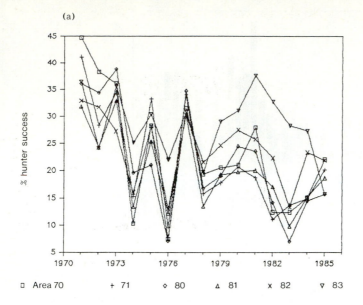

(b)

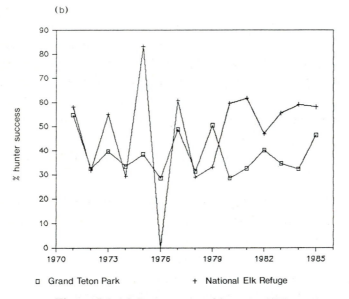

Figure 7.2 (a) Percentage of hunters killing elk on six hunt areas on National Forest lands in the Jackson herd unit, 1971–85. (b) Percent kill success for hunters in Grand Teton National Park and on the National Elk Refuge (area 77), 1971–85. Due to a mild autumn, no elk migrated onto the Refuge in 1976 to secure a harvest.

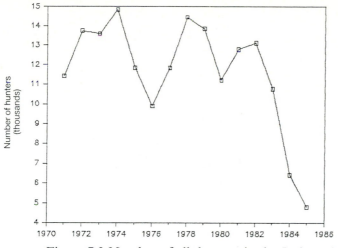

Figure 7.3 Number of elk hunters in the Jackson herd unit, 1971–85.

This apparently muddled Department efforts to manipulate harvest strategies, as evidenced by the following statement (Straley, Roby & Johnson, 1984):

> Seasons in the past several years have been designed to reduce the Refuge herd segment below the 7,500 elk allowed, thus allowing time to increase those elk not associated with the two national parks. It is now possible to build those populations back to higher levels and still maintain reduced hunting pressure on the Park herd segments. These objectives will be very difficult to achieve before the Refuge and Park populations are back to former levels, consequently, a reduced but adequate harvest will still have to be taken off the Grand Teton National Park herd segment.

Consequently, the Game and Fish Department proposed that harvest objectives be lowered for the Jackson herd, seasons be shortened, and in some areas, harvests be restricted to branch-antlered bulls (Table 7.1).

Planning can enhance the wildlife management process by enforcing clearer thinking of management alternatives and by articulating the reasons for management objectives and strategies to the public and other agencies (Crowe, 1983). Yet, there can be danger in overplanning (D. Lockman, Wyoming Game and Fish Department, pers. comm.). Good planning must maintain flexibility to change management objectives when improved knowledge becomes available or when desires of the public change. But at the same time, part of the purpose for the wildlife planning process is to reduce 'knee-jerk' reactions to public pressures. There must be an optimal middle ground for informed adjustments of management objectives.

Table 7.1. *Elk hunting regulations for Jackson elk herd (see map at Figure 3.22).*

Hunt area	1986	1985	1984	1983
Area 70 Buffalo Fork				
Seasons	10 Sep–31 Oct	10 Sep–31 Oct	10 Sep–2 Nov	10 Sep–15 Nov; 29 Oct–4 Nov
Restrictions	Mature bulls only	Mature bulls only	Antlered elk	Antlered elk; any elk
Permits	None	50 spikes excluded	50 any elk	
Area 71 Pacific Creek				
Seasons	10 Sep–31 Oct	10 Sep–31 Oct	10 Sep–11 Nov	10 Sep–28 Oct; 29 Oct–30 Nov
Restrictions	Mature bulls only	Mature bulls only	Antlered elk	Antlered elk; any elk
Permit type 1	None	50 spikes excluded	50 any elk	
Permit type 2	100 spikes excluded	None	50 any elk	
Type 2 season	15 Oct–14 Nov		27 Oct–11 Nov	
Area 72 Berry Creek				
	Closed	Closed	Closed	Closed
Area 74 Ditch Creek				
Seasons	Combined with area 81	Combined with area 81	Combined with area 81	10 Sep–15 Nov; 24 Oct–4 Nov
Restrictions				Antlered elk; any elk
Area 75 Riverbottom				
Seasons	25 Oct–14 Nov	26 Oct–15 Nov	27 Oct–11 Nov	29 Oct–30 Nov
Restrictions	Any elk by permit only	Any elk by permit only	Any elk by permit only	33 antlerless and 12 any elk permits at a time
Permits	90 each week	90 each week	90 each week	45 permits/time period
Area 76 Blacktail/Hayfields				
Seasons	25 Oct–14 Nov	26 Oct–15 Nov	27 Oct–11 Nov	29 Oct–30 Nov
Restrictions	Any elk by permit only	Any elk by permit only	Any elk by permit only	Antlerless elk by permit only
Permits	160 each week, of which 90 for only Blacktail Butte	160 each week, of which 90 for only Blacktail Butte	160 each week for only Blacktail Butte	135 each period, of which 60 (18 any elk) Blacktail Butte only
Area 77 National Elk Refuge				
Seasons	25 Oct–14 Nov	26 Oct–15 Nov	27 Oct–16 Nov	29 Oct–30 Nov
Restrictions	Any elk by permit only	Any elk by permit only	Any elk by permit only	10 any, 30 antlerless elk by permit only
Permits	120 each week, max. 40 hunters at a time	120 each week, max. 40 hunters at a time	120 each week, max. 40 hunters at a time	120/week, max. 40/day
Area 78 Wilson				
Seasons	Closed	Closed	15 Nov–15 Dec	15 Nov–15 Dec
Restrictions			Archery only	Archery only
Area 79 Teton Park				
Seasons	25 Oct–14 Nov	26 Oct–15 Nov	27 Oct–11 Nov	29 Oct–30 Nov
Restrictions	Any elk by permit only	Any elk by permit only	Any elk by permit only	Any elk by permit only
Permit type 1	1500 any elk	1000 any elk	1000 any elk	1500 any elk
Permit type 2	None	None	800 any elk	1000 any elk
Type 2 season			10 Nov–11 Nov	12 Nov–30 Nov
Area 80 Sheep Creek				
Seasons	10 Sep–31 Oct	10 Sep–31 Oct	10 Sep–2 Nov	10 Sep–4 Nov
Restrictions	Mature bulls only	Mature bulls only	Antlered elk	Antlered elk
Area 81 Spread Creek				
Seasons	10 Sep–31 Oct	10 Sep–31 Oct	10 Sep–2 Nov	10 Sep–15 Nov; 24 Oct–4 Nov
Restrictions	Mature bulls only	Mature bulls only	Antlered elk	Antlered elk
Permits	100 antlerless elk	100 spikes excluded	200 any elk	
Area 82 Crystal Creek				
Seasons	10 Sep–31 Oct	10 Sep–31 Oct	10 Sep–2 Nov	10 Sep–31 Oct; 10 Sep–15 Nov
Restrictions	Mature bulls only	Mature bulls only	Antlered elk	Antlered elk; permit only
Permits	100 antlerless elk	100 spikes excluded	200 any elk	250 any elk
Area 83 Fish Creek				
Seasons	1 Oct–31 Oct	1 Oct–31 Oct	1 Oct–31 Oct	1 Oct–31 Oct
Restrictions	Mature bulls only	Mature bulls only	Antlered elk	Antlered elk
Permits	100 antlerless elk	100 spikes excluded	200 any elk	300 any elk

1982	1981	1980	1979
10 Sep–29 Oct; 30 Oct–15 Nov Antlered elk; any elk	10 Sep–30 Oct; 31 Oct–15 Nov; 16 Nov–6 Dec Antlered elk; any elk; Any elk in part of area	10 Sep–17 Oct; 18 Oct–15 Nov; 16 Nov–7 Dec Antlered elk; any elk Any elk in part of area	10 Sep–19 Oct; 20 Oct–15 Nov, 1–9 Dec Antlered elk; any elk
10 Sep–29 Oct; 30 Oct–31 Dec Antlered elk; any elk	10 Sep–30 Oct; 31 Oct–31 Dec Antlered elk; any elk	10 Sep–17 Oct; 18 Oct–31 Dec Antlered elk; any elk	10 Sep–19 Oct; 20 Oct–31 Dec Antlered elk; any elk
Closed	Closed	Closed	Closed
10 Sep–24 Oct; 25 Oct–5 Dec Antlered; antlerless elk	10 Sep–20 Oct; 21 Oct–6 Dec Antlered; Antlerless elk	10 Sep–17 Oct; 18 Oct–7 Dec Antlered; antlerless elk	10 Sep–19 Oct; 20 Oct–9 Dec Antlered; antlerless
30 Oct–5 Dec 54 antlerless and 6 any elk permits at a time 60 permits/time period	31 Oct–6 Dec 30 antlerless and 6 any elk permits at a time 36 permits/time period	25 Oct–7 Dec 25 antlerless and 5 any elk permits at a time 30 permits/time period	27 Oct–9 Dec Antlerless elk by permit only 20 permits/week
30 Oct–5 Dec Antlerless elk by permit only 150 each period, of which 60 (18 any elk) Blacktail Butte only	31 Oct–6 Dec Antlerless elk by permit only 72 each period, of which 36 (6 any elk) Blacktail Butte only	25 Oct–7 Dec Antlerless elk by permit only 70 each period, of which 25 (10 any elk) Blacktail Butte only	27 Oct–9 Dec Antlerless elk by permit only 60/week, 20 for Blacktail Butte, 40 for Hayfields
30 Oct–5 Dec Any elk until migration, then only antlerless 150/week, max. 50/day	31 Oct–6 Dec Any elk until migration, then only antlerless 120/week, max. 40/day	25 Oct–7 Dec Any elk until migration, then only antlerless 120/week, max. 40/day	20 Oct–2 Nov; 3 Nov–9 Dec Any elk; antlerless elk by permit only 120/week, max. 40/day
15 Nov–5 Dec Archery only	15 Nov–15 Dec Antlerless elk	15 Nov–15 Dec Antlerless elk	15 Nov–15 Dec Antlerless elk
30 Oct–5 Dec Any elk by permit only 1500 any elk 1500 any elk 13 Nov–5 Dec	31 Oct–6 Dec Any elk by permit only 1500 any elk 1000 any elk 14 Nov–6 Dec	25 Oct–7 Dec Any elk by permit only 1500 any elk 1000 any elk 8 Nov–7 Dec	27 Oct–9 Dec Any elk by permit only 1500 any elk 1000 any elk 10 Nov–9 Dec
10 Sep–24 Oct; 25 Oct–10 Nov; 11 Nov–5 Dec Antlered; any; antlerless Noon opening 3rd season	10 Sep–20 Oct; 21 Oct–6 Nov; 7 Nov–6 Dec Antlered; any; antlerless Noon opening 3rd season	10 Sep–17 Oct; 18 Oct–31 Oct; 1 Nov–7 Dec Antlered; any; antlerless Noon opening 3rd season	10 Sep–19 Oct; 20 Oct–2 Nov; 3 Nov–9 Dec Antlered; any; antlefless Noon opening 3rd season
10 Sep–24 Oct; 25 Oct–15 Nov; 16 Nov–5 Dec Antlered; any; antlerless	10 Sep–20 Oct; 21 Oct–15 Nov; 16 Nov–6 Dec Antlered; any; antlerless	10 Sep–17 Oct; 18 Oct–15 Nov; 16 Nov–7 Dec Antlered; any; antlerless In part of area	10 Sep–19 Oct; 20 Oct–15 Nov; 16 Nov–9 Dec Antlered; any; antlerless
10 Sep–31 Oct; 10 Sep–15 Nov Antlered elk; permit only 300 any elk	10 Sep–31 Oct; 1 Oct–15 Nov Antlered elk; permit only 300 any elk	10 Sep–15 Nov; 18 Oct–15 Nov Antlered elk; permit only 300 any elk	10 Sep–19 Oct; 20 Oct–15 Nov Antlered elk 300 any elk permits for late season only
1 Oct–31 Oct Antlered elk 350 any elk	1 Oct–31 Oct Antlered elk 300 any elk	1 Oct–31 Oct; 15–31 Oct Antlered elk 300 any elk permits for late season only	1–14 Oct; 15–31 Oct Antlered elk 250 any elk permits for late season only

Table 7.1 *(cont.)*

Hunt area	1978	1977	1976	1975
Area 70 Buffalo Fork				
Seasons	10 Sep–20 Oct; 21 Oct–15 Nov, 1–15 Dec	10 Sep–21 Oct; 22 Oct–13 Nov, 1–15 Dec	10 Sep–22 Oct; 23 Oct–7 Nov, 1–7 Dec	10 Sep–20 Oct; 21 Oct–9 Nov, 1–7 Dec
Restrictions	Antlered elk; any elk	Antlered elk; any elk	Antlered elk; any elk	Antlered elk; any elk
Permits			N of Pac Ck, 8–30 Nov, any	
Area 71 Pacific Creek				
Seasons	10 Sep–20 Oct; 21 Oct–31 Dec	10 Sep–21 Oct; 22 Oct–31 Dec	10 Sep–22 Oct; 23 Oct–30 Nov	10 Sep–20 Oct; 21 Oct–30 Nov
Restrictions	Antlered elk; any elk	Antlered elk; any elk	Antlered elk; any elk	Antlered elk; any elk
Area 72 Berry Creek				
	Closed	Closed	Closed	Closed
Area 74 Ditch Creek				
Seasons	10 Sep–20 Oct; 21 Oct–3 Dec	10 Sep–28 Oct	10 Sep–22 Oct	10 Sep–20 Oct; 21–31 Oct
Restrictions	Antlered; antlerless	Antlered elk	Antlered elk	Antlered elk; any elk
Area 75 Riverbottom				
Seasons	28 Oct–3 Dec	29 Oct–30 Nov		Closed
Restrictions	Antlerless elk	Antlerless elk		
Permits	20 permits/week	20 permits/week		
Area 76 Blacktail/Hayfields				
Seasons	28 Oct–3 Dec	29 Oct–30 Nov	23 Oct–30 Nov	1–30 Nov
Restrictions	Antlerless elk by permit only	Antlerless elk by permit only	Antlerless elk by permit only	Any elk by permit only
Permits	60/week, 20 for Blacktail Butte, 40 for Hayfields	60/week, 20 for Blacktail Butte, 40 for Hayfields	60/week, 20 for Blacktail Butte, 40 for Hayfields	60/week, 10 for Blacktail Butte, 50 for Hayfields
Area 77 National Elk Refuge				
Seasons	28 Oct–8 Dec	29 Oct–17 Nov	29 Oct–17 Nov	25 Oct–13 Nov
Restrictions	Any elk by permit only	Any elk by permit only	Any elk by permit only	Any elk by permit only
Permits	120/week, max. 40/day	90/week, max. 30/day	80/week, max. 30/day	60/week, max. 30/day
Area 78 Wilson				
Seasons	15 Nov–15 Dec	15 Nov–15 Dec	15 Nov–15 Dec	15 Nov–15 Dec
Restrictions	Antlerless elk	Antlerless elk	Antlerless elk	Any elk
Area 79 Teton Park				
Seasons	28 Oct–3 Dec	29 Oct–30 Nov	23 Oct–30 Nov	1–30 Nov
Restrictions	Any elk by permit only	Any elk by permit only	Any elk by permit only	Any elk by permit only
Permit type 1	1500 any elk	1500 any elk	1500 any elk	1000 any elk
Permit type 2	1000 any elk	1000 any elk	1000 any elk	1500 any elk
Type 2 season	11 Nov–3 Dec	5 Nov–30 Nov	6 Nov–30 Nov	8–30 Nov
Area 80 Sheep Creek				
Seasons	10 Sep–27 Oct; 28 Oct–8 Dec	10 Sep–21 Oct; 22 Oct–30 Nov	10 Sep–22 Oct; 23 Oct–14 Nov; 15–30 Nov	10 Sep–20 Oct; 21–24 Oct; 25 Oct–30 Nov
Restrictions	Antlered; antlerless. Noon opening 2nd season	Antlered; antlerless. Noon opening 2nd season	Antlered; antlerless. Noon opening 3rd season	Antlered; any; antlerless. Noon opening 3rd season
Area 81 Spread Creek				
Seasons	10 Sep–20 Oct; 21 Oct–15 Nov; 16 Nov–3 Dec	10 Sep–21 Oct; 22 Oct–13 Nov; 14 Nov–15 Dec	10 Sep–22 Oct; 23 Oct–7 Nov	10 Sep–20 Oct; 21 Oct–9 Nov
Restrictions	Antlered; any; antlerless	Antlered; any; antlerless	Antlered elk; any elk	Antlered elk; any elk
Area 82 Crystal Creek				
Seasons	10 Sep–20 Oct; 21 Oct–15 Nov	10 Sep–21 Oct; 22 Oct–13 Nov	10 Sep–22 Oct; 23 Oct–7 Nov	10 Sep–20 Oct; 21 Oct–9 Nov
Restrictions	Antlered elk	Antlered elk	Antlered elk; any elk	Antlered elk; any elk
Permits	300 any elk permits for late season only	300 any elk permits for late season only		
Area 83 Fish Creek				
Seasons	1–14 Oct; 15–31 Oct	1–14 Oct; 15–31 Oct	10 Sep–22 Oct; 23–31 Oct	10 Sep–20 Oct; 21 Oct–9 Nov
Restrictions	Antlered elk	Antlered elk	Antlered elk; any elk	Antlered elk; any elk
Permits	250 any elk permits for late season only	250 any elk permits for late season only		

1974	1973	1972	1971
10 Sep–14 Oct; 15 Oct–3 Nov Mature bull; any elk but spikes excluded	10 Sep–14 Oct; 15 Oct–2 Dec; 15–23 Dec Mature bull; any, no spikes; any elk in Buffalo Fk	10–30 Sep; 1 Oct–15 Nov; 15–31 Dec Mature bull; Any, no spikes; any elk in Buffalo Fk	10–30 Sep; 1 Oct–10 Nov; 15–31 Dec Antlered; any; archery
10 Sep–14 Oct; 15 Oct–1 Dec Mature bulls; Any elk but spikes excluded	10 Sep–14 Oct; 15 Oct–2 Dec Mature bulls; Any elk but spikes excluded	10–30 Sep; 1 Oct–30 Nov Mature bulls; Any elk but spikes excluded	10–30 Sep; 1 Oct–30 Nov Antlered elk; any elk
Closed	Closed	Closed	Closed
10 Sep–14 Oct; 15 Oct–1 Nov Mature bull; any elk but spikes excluded	10 Sep–14 Oct; 15–31 Oct Mature bulls; any elk but spikes excluded	10–30 Sep; 1–20 Oct Mature bulls; any elk but spikes excluded	10–30 Sep; 1 Oct–30 Nov Antlered elk; any elk
Closed	3 Nov–2 Dec Any elk, spikes excluded, by permit only 50 permits per week	21 Oct–8 Dec Any elk by permit only 50 permits per week	23 Oct–30 Nov Any elk by permit only 50 permits per week
2 Nov–1 Dec Any elk by permit only 60/week, 10 for Blacktail Butte, 50 for Hayfields	3 Nov–2 Dec Any elk, spikes excluded, by permit only 50 permits per week	21 Oct–8 Dec Any elk by permit only 50 permits per week	23 Oct–30 Nov Any elk by permit only 50 permits per week
25 Oct–13 Nov Any elk, spikes excluded, by permit only 80/week, max. 40/day	27 Oct–9 Nov Antlerless elk by permit only 80/week, max. 40/day	21 Oct–10 Nov Antlerless elk by permit only 80/week, max. 40/day	16 Oct–5 Nov Antlerless elk by permit only 60/week
1 Nov–31 Dec Any elk spikes excluded	10 Sep–14 Oct; 15 Oct–31 Dec Mature bulls; any elk, spikes excluded	10–30 Sep; 1 Oct–31 Dec Mature bulls; any elk, spikes excluded	10–30 Sep; 1 Oct–31 Dec Antlered elk; any elk
2 Nov–1 Dec Any elk, spikes excluded, by permit only 1000 any but spikes 1500 any but spikes 9 Nov–1 Dec	3 Nov–2 Dec Any elk, spikes excluded, by permit only 1000 any but spikes 1500 any but spikes 10 Nov–1 Dec	21 Oct–8 Dec Any elk by permit only 1000 any elk permits 1500 any elk permits 11 Nov–8 Dec	23 Oct–30 Nov Any elk by permit only 1200 any elk permits 1300 any elk permits 6–30 Nov
10 Sep–14 Oct; 15 Oct–1 Nov; 2–15 Nov No spikes Mature bulls; any; any; Noon opening 3rd season	10 Sep–14 Oct; 15–26 Oct; 27 Oct–11 Nov No spikes Mature bulls; any; any; Noon opening 3rd season	10–30 Sep; 1–31 Oct; 1–20 Nov No spikes Mature bulls; any; any; Noon opening 3rd season	10–30 Sep; 1–31 Oct Antlered elk; any elk
10 Sep–14 Oct; 15 Oct–17 Nov Mature bulls; any elk, spikes excluded	10 Sep–14 Oct; 15 Oct–30 Nov; 15–23 Dec Mature bulls; any, no spikes; any elk in Buffalo Fk	10–30 Sep; 1 Oct–30 Nov; 15–31 Dec Mature bulls; any, no spikes; any elk in Buffalo Fk	10–30 Sep; 1 Oct–20 Nov; 15–31 Dec Mature bulls; any, no spikes; mature bull, any archery
10 Sep–14 Oct; 15 Oct–17 Nov Mature bulls; any elk, spikes excluded	10 Sep–14 Oct; 15 Oct–30 Nov Mature bulls; any elk, spikes excluded	10–30 Sep; 1 Oct–30 Nov Mature bulls; any elk, spikes excluded	10–30 Sep; 1 Oct–20 Nov Any elk, spikes excluded; antlered elk
10 Sep–14 Oct; 15 Oct–17 Nov Mature bulls; any elk, spikes excluded	10 Sep–14 Oct; 15 Oct–30 Nov Mature bulls; any elk, spikes excluded	10–30 Sep; 1 Oct–30 Nov Mature bulls; any elk, spikes excluded	10–30 Sep; 1 Oct–20 Nov Any elk, spikes excluded; antlered elk

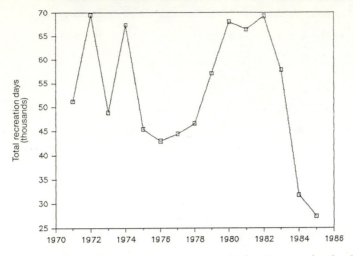

Figure 7.4 Total recreation days for hunters in the Jackson herd unit, 1971–85. This is equivalent to the number of successful hunters times the average number of days required to kill an elk.

Public resistance to heavy kills from the Jackson elk herd ultimately influences subsequent harvests. In fact, Beetle (1979) claimed that hunting has been unsuccessful at maintaining elk numbers at or below 'carrying capacity'. Overlooking a semantic argument over definitions for carrying capacity (Caughley, 1979; MacNab, 1985), Beetle's challenge is interesting. He suggests that kills have never been adequate because high kills are 'always followed by a reactionary underkill'. He offers as evidence that 6000 elk were killed in 1935, but only 2000 the following year. Likewise, a kill of 5500 in 1943 was followed by only 1170 in 1944, and a kill of 6000 in 1951 was followed by 3000 in 1952.

Is it true that large kills are followed by a reactionary 'underharvest'? This suggests that annual harvest statistics should show a negative autocorrelation coefficient at a short lag. In Figure 7.5 I present the autocorrelation function for first-order differences of hunter harvest estimates from 1950–84. I have not included earlier years because procedures for estimating harvest do not permit a consistent comparison with later years when more careful procedures have been employed. We see that the autocorrelation for the first lag is significantly negative, which supports Beetle's contention. This pattern appears to be a response by hunters rather than by the elk population because the pattern does not appear in the autocorrelation of Refuge census statistics. Whether the apparent reactionary reduction in harvest following a heavy one implies underharvest is not clear, but it does appear that public pressure to sustain the Jackson elk herd may interfere with harvest objectives.

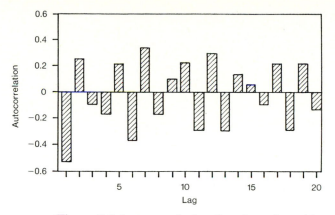

Figure 7.5 Autocorrelation function of total hunter kill, 1950–84, after first-order differencing. Only the first lag shows a statistically significant ($P < 0.01$) autocorrelation.

Hunter harvest

In Figure 7.1 I present the total harvest from the Jackson elk herd relative to harvest objectives. Reduced harvests in 1984 and 1985 were obtained by (1) manipulating length and timing of hunting seasons, (2) by restricting number of permits issued for limited quota hunts, for example, in Grand Teton National Park, and (3) by restricting the harvest in some areas to antlered, or even branch-antlered, bulls only. In Table 7.1 I have listed seasons and regulations for 1971–86, showing regulations spanning a range in harvest from 1981 when a large harvest of 4290 elk occurred, and those for 1985, when a much-reduced harvest of 1368 elk was obtained.

Hunter kill is distributed over the herd unit by allocating different regulations for the various hunt areas described in Chapter 3. Hunter success varies considerably among hunt areas (Figure 7.2). Hunters quickly learn which areas provide the highest hunter success by personal experience or from information provided by the Wyoming Game and Fish Department (e.g., the *Annual Report of Big Game Harvest*). Such knowledge is not counter-productive for management; rather, it helps to favorably distribute hunters who are free to hunt in several hunt areas. That this occurs is supported by the significant correlation between hunter success rates and the change in total recreation days to the following year over all non-Park hunt areas, standardized for each year ($r = 0.227$, $n = 91$, $P = 0.03$).

Hunter recreation days and the mean number of days that hunters spend to kill an elk are illustrated in Figures 7.4 and 7.6 respectively. As would be expected, hunter success and the number of days required to harvest an elk are inversely correlated ($r = -0.832$, $n = 13$, $P < 0.01$, Figure 7.7).

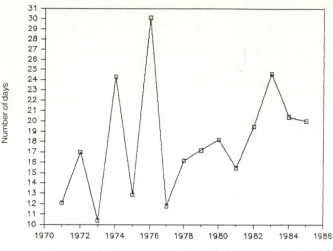

Figure 7.6 Average number of days required to kill an elk in the Jackson herd unit.

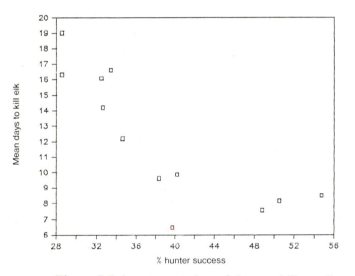

Figure 7.7 Average number of days to kill an elk versus percentage hunter success in Grand Teton National Park, 1971–83.

Harvest strategies

Despite the control over harvest possible through adjustments in hunting seasons and regulations, the elk harvest still varies considerably. Migration timing is one factor which reduces the effectiveness of timing hunting seasons to manipulate harvests. As seen in Chapter 2, heavy snowfall precipitates early migration, and if seasons are open along migration routes, the kill can be substantial (Anderson, 1958). Conversely, during autumns of little snowfall, such as in 1976 and 1980, migration may be very late and harvests may be extraordinarily small. This pattern is demonstrated by the correlation between hunter success and maximum depth of snow on the ground in November at Moran ($r = 0.55$, $n = 13$, $P < 0.05$). Because of the unpredictability of weather, therefore, using the time or length of hunting season to manipulate elk harvests may have unpredictable effectiveness. This is a serious management concern. Despite extended seasons for elk on National Forest lands during 1975–83, total kill from these areas averaged only 2321 elk. During the same period, an average of 884 elk were taken during the Park and Refuge hunts, or 28% of the total harvest. Without the Park and Refuge hunts to ensure harvest, the Wyoming Game and Fish Department is concerned that it may not be able to secure adequate cull of elk to keep the herd within herd objectives (Thomas, Crowe & Kruckenberg, 1984). As shown in Chapter 3, harvests from the Park and Refuge constitute a higher fraction of the total kill during years of low total kill.

Hunter success averages higher for Park and Refuge hunters in most years (Figure 7.2), and total kill is highly correlated with the number of permits issued. Hunts in the Park and on the Refuge are usually scheduled to occur during early elk migration and are presumed to cull largely elk that summer in the Park. Migration data (see Chapter 2 and Smith, 1985) support this presumption.

Due to legal restrictions, elk that summer west of the Snake River in Grand Teton National Park and in Yellowstone National Park are inaccessible to hunters until they migrate out of these areas. For the most part, harvest on the Grand Teton Park herd segment must occur during hunts in Grand Teton National Park and on the National Elk Refuge. Elk coming from Yellowstone National Park may be harvested on Forest lands during migration, but may also be susceptible to harvest in Grand Teton National Park and on the Refuge, particularly if migration occurs early.

Closer control on harvests can be obtained by limited quota permits for specific hunt areas. Even though hunter success may vary, the maximum harvest cannot exceed the number of permits issued. Limited quota permits are used in the Park and Refuge hunts and extensively in other parts of Wyoming, but guides and outfitters have strongly resisted using this approach for the Jackson elk herd. This is understandable because a client may plan a wilderness elk hunt months if not years in advance, and some guides may be booked up well in advance of when limited quota permits are issued, typically in July. The inability to predict whether a client will be randomly selected to receive a limited quota permit is understandably undesirable for guides who may not be able to secure a replacement client in time for a September hunt.

A compromise solution might be to ensure each licensed guide a limited number, say five, of limited quota permits (Ridge Taylor, outfitter and guide, pers. comm.). These could be sold to hunters wishing to hunt with the guide so that security in clientele bookings could be assured. This would allow much closer control on harvests while not devastating the guiding industry. However, such a proposal will meet resistance from hunters who do not use guiding services, because it offers an unfair advantage to guides.

Harvests of elk during the late 1950s and early 1960s apparently drastically reduced a migratory segment crossing Highway 26/287 between Togwotee Lodge and Togwotee Pass (see Chapter 2). Assuming that this has resulted in reduced use of intact summer ranges previously occupied by these elk, it seems to me desirable to attempt to rebuild this segment of the population by reduced harvest over a number of years. If managers wanted to rebuild this herd segment, shortening the hunting season in the Teton Wilderness may be ineffective, because migration along this route is usually among the earliest for any herd segment. A later season may help to build this herd segment, but this would be undesirable for guides and outfitters because snow accumulation later in the season makes for difficult and dangerous back-country hunts.

Initiation of a limited-quota season in the eastern half of area 70 (Figure 3.22) probably would be a more effective way of increasing the number of elk in this area. This could be best accomplished by subdividing area 70 into two new hunt areas. Continuing the current restriction for harvesting only branch-antlered bulls may achieve the same objective, so long as bulls are not excessively harvested. Removal of bulls to the point that spikes do most of the breeding can result in cows being bred during their second estrus. Later born calves may not be large enough by winter to survive (Prothero, Spillett & Balph, 1979).

Other than area 70, the current hunt areas appear to be reasonably effective at accomplishing the harvest distribution for which they were designed.

Cole (1969) had predicted that by increasing focus on culls in Grand Teton National Park and the National Elk Refuge, within a decade it should have been possible to restore historical migrations in eastern portions of the Teton Wilderness which supported elk historically (Anderson, 1958). However, inspection of harvest regulations (Table 7.1) and associated kills (figures 3.26–3.36) do not show a continued harvest pattern consistent with restoring migration patterns. Current restrictions on harvest appear to distribute kills more appropriately. Yet, as the herd builds, it will be necessary to secure greater culls from western segments of the herd, for example, from Grand Teton National Park and the National Elk Refuge, if historical migrations are to be restored.

Damage costs

Because wildlife in Wyoming is legally the property of the state, in 1939 the state legislature enacted Wyoming Statute Chapter 65, Section 45 requiring the Wyoming Game and Fish Department to compensate property owners for damages due to wildlife. The current Wyoming Statute in force is 23–1–901. Most damage by elk in Jackson Hole involves consumption of hay during winter. Damage claims are paid from the Department's damage fund, which is sustained by fees collected from the sale of nonresident big game licenses (Crowe, 1981).

Payments for elk damage within the Jackson herd unit from 1974 through 1980 totalled only $1270 and occurred only in 1978 ($1170) and 1980 ($100). This averages only $181 each year or approximately $0.01 per elk, compared to $0.63 per elk for other elk herds in Wyoming (Crowe, 1981). However, wildlife damage claims state-wide have increased markedly in recent years. For example, elk damage claims in the Jackson herd unit in 1983 and 1984 totalled $9416. All of this was for hay loss except for $75 paid to one landowner for a tree destroyed by elk on her lawn.

State contributions to feeding

The Wyoming Game and Fish Department currently buys half the feed for the National Elk Refuge feeding program. In addition, the Department solely supports the Patrol Cabin, Fish Creek and Alkali Creek feedgrounds in the upper Gros Ventre River valley. Costs to the state for the Gros Ventre feedgrounds run as high as $75 000 annually, and nearly $200 000 for the entire herd unit feeding program.

The Wyoming Game and Fish Commission's official policy on supplemental feeding of big game is:

> To keep wildlife populations in balance with local forage supplies and commensurate with conflicting uses, and manage on the basis of available forage, resorting to artificial feeding only in previously approved on-going programs or on an emergency case-by-case basis (Wyoming Game and Fish Department, 1983).

The Wyoming Game and Fish Department recognized as a problem that 'Some wildlife users demand high cost projects to maintain populations of some species above the natural carrying capacity of their habitat' (Wyoming Game and Fish Department, 1983). Their 'strategic plan' suggests two strategies for solving the problem: (a) 'emphasize the high cost of artificial animal production systems and the possible biological and sociological detriments of these projects,' and (b) 'develop and implement methods whereby the users of high cost production projects pay their fair share of the management costs' (Wyoming Game and Fish Department, 1983).

Currently, the Wyoming Game and Fish Department's contribution to feeding for the Jackson elk herd is supported from funds secured through sale of hunting licenses. Obviously, the Department desires to allocate such feeding monies so as to best enhance returns to hunters that provide the funds. This desire is unquestionably an important consideration in Department policy toward management for the Jackson elk herd. Maximizing returns for hunters implies managing for maximum sustained yield, or at least close to this level. Evidence reviewed in Chapter 3 suggests that, on average, harvesting programs during the past 20 years appear to have been quite close to maximum sustained yield.

The herd reduction program in Grand Teton National Park appears to create some policy conflicts for the Wyoming Game and Fish Department. The Department's desire to maintain high yield from the Jackson elk herd may be viewed to be in conflict with an agreement which the Department signs when filing the joint recommendation with the National Park Service to institute the annual herd reduction program. At least since 1964, the Wyoming Game and Fish Department has agreed to 'long-range objectives of reducing the *need* to kill large numbers of elk within Grand Teton National Park' (Grand Teton National Park, 1985).

The Department finds it *necessary* to cull elk in the Park because it is impossible to secure adequate harvest on this herd segment elsewhere. Necessity for a cull is based upon the premise that without a cull the Park herd segment will become a larger proportion of the National Elk Refuge herd. Partly for financial reasons, the Refuge herd is constrained to an objective of no more than 7500 elk. It is thought that increasing harvests outside the Park to keep the Refuge herd less than 7500 will eventually result in fewer elk available for hunter harvest. This concern is justified, although experimental proof does not exist.

The yield curve presented in Chapter 3 (Figure 3.33) suggests that the elk population summering in the Park will increase by approximately 30–50% before reaching carrying capacity. If indeed the Refuge quota is maintained at 7500 elk this means that a larger proportion of the Refuge herd must be elk summering within the boundaries of Grand Teton National Park. Because density-dependent survival of calves and yearlings contributes to the attainment of carrying capacity (see Chapter 3), hunter harvest from the Jackson elk herd will be reduced if the Park herd is permitted to increase without harvest.

As a consequence, the Wyoming Game and Fish Department feels that increasing the fraction of Park elk on the National Elk Refuge will reduce the commitment to support feeding operations with State funds which come largely from the sale of hunting licenses (Dexter, 1984). And it seems unlikely that the National Park Service would support an artificial feeding program, despite potential benefits to nonconsumptive users in Grand Teton National Park.

Population modeling

The Game Division of the Wyoming Game and Fish Department encourages biologists in each district to model big game populations. The computer model is a demographic projection model of a series developed initially by J. Gross (1969).

Inputs to the model include calf/cow ratios, harvest by sex and age group, pre- and post-season sex and age classifications, approximate overwinter survival rates, and an estimate of population size. With the best estimates for these parameters, a population trajectory is simulated. If the behavior of the model seems unreasonable, that is, if it forecasts a dramatic rise or fall in population size, estimated parameters are adjusted until plausible dynamics are observed. Manipulation of estimated parameters is iterative, and the biologist manipulates those estimates with which he is least confident. The ultimate objective is to find a combination of demographic parameters which matches the previously observed behavior of the population and harvest estimates (Pojar, 1981).

Attempts to prepare such a model for the Jackson elk herd have been largely unsuccessful. Although reliable estimates exist for several parameters, for some unknown reason, population projections typically forecast dramatic declines in the elk herd at observed harvest levels (G. Roby, pers. comm.). Reasonable projections can be obtained, however, by assuming 2000 more elk in the herd than are known to winter in the valley. The implication is that many elk winter on ranges away from feedgrounds in the Jackson herd unit, or move outside the herd unit boundaries. Yet these animals are within the herd unit during hunting season and supplement the harvestable population. The fact that feedground attendance is so highly correlated with winter weather (Chapter 3) suggests that numerous elk may be wintering away from feedgrounds during most winters.

Demographic projection models may be criticized on numerous grounds. First, parameter estimates often possess large standard errors, and the iterative procedure employed to make the model 'work' implies that an infinite number of possible 'acceptable' solutions exist. Furthermore, a demographic model does not directly incorporate interactions between elk and habitat. On the one hand, it might be argued that reliable field estimates of survival and recruitment provide all the information that is necessary to characterize fluctuations in a population. Observed variation in parameter estimates will mirror habitat interaction, for example, density dependence in recruitment rates will be observed in the field and therefore input will ultimately reflect habitat interactions. Yet, due to time lags in habitat response, dynamics of projection models can be misleading.

Another critical weakness of projection models is that they cannot predict an optimum or maximum sustained yield harvest level without incorporating explicit density-dependent functions for recruitment and survival (Mendelssohn, 1976). Yet, measuring details of density dependence requires extensive field study that may not be practical where substantial population manipulations are incompatible with management objectives.

Nevertheless, I think that the modeling efforts have been very useful in providing field biologists with (1) improved understanding of demographic interactions and consequences to population dynamics, (2) a tool for exploring possible consequences of changes in management tactics and harvest policy, and (3) an appreciation for the value of obtaining good estimates for population parameters. When combined with a solid appreciation for the critical role of habitat interactions in determining variation in animal recruitment and survival, population projection modeling can certainly improve understanding of population processes and, therefore, is valuable. I think that attempts to fit a demographic model to the Jackson elk herd data offer an important message: there are probably more elk in the harvestable population than are currently being counted in Jackson Hole.

Summary

1. Population size for the Jackson elk herd is determined by the Wyoming Game and Fish Department which regulates by harvest. Herd objectives are to maintain no more that 7500 elk on the National Elk Refuge and a total of 2182 elk at the three State feedgrounds in the upper Gros Ventre valley.
2. Large hunter harvests of elk in the Jackson herd unit are usually followed by substantially smaller harvests, partly due to response to public pressure to protect the elk herd from overexploitation.
3. To consistently achieve an elk harvest large enough to maintain the elk herd within desired limits, it is necessary to cull elk in Grand Teton National Park. To restore and maintain the historical distribution of elk within the Jackson herd unit, substantial harvests of elk from the Park and the Refuge will be necessary.

4. The State of Wyoming is financially responsible when elk cause damage to private property. This damage usually entails consumption of private hay, totalling approximately $5000 annually in the Jackson elk herd unit.
5. Demographic modeling by the Wyoming Game and Fish Department shows that attempts to census elk in the Jackson herd leads to underestimates of actual population size.

Chapter 8.

Management of elk in Grand Teton National Park

The policy of the National Park Service is to maintain lands in pristine condition, as near as possible to that existing before European man first settled the area (Houston, 1982). Since certain components of the ecosystem have been altered by man (e.g., wolves [*Canis lupus*] have been eliminated and natural fires have been suppressed), most Parks are actually managed by an unstated policy of non-interference. An important rationale for this policy is to maintain areas as ecological baselines. Without self-sustaining natural ecosystems, we have no basis for assessing causes of ecological change in other areas (Sinclair, 1983). Because of historical constraints surrounding the establishment of Grand Teton National Park, there are two major exceptions to the general policy of non-interference in the Park: (1) livestock grazing, and (2) the elk reduction program. In this chapter I discuss management of the Jackson elk herd by the National Park Service, with an emphasis on the controversial hunt that occurs within the boundaries of Grand Teton National Park.

The boundaries of the Jackson elk herd encompass the entire Grand Teton National Park (1264 km²). Many elk summer within the Park and many migrate through the Park en route to and from the National Elk Refuge. Elk from the Jackson herd also range over approximately 800 km² in southern portions of Yellowstone National Park during the summer. For the most part, those animals are undisturbed whilst in these wilderness portions of Yellowstone National Park.

Law 81–787 and management of Grand Teton National Park

Formation of Grand Teton National Park has a complex history (Righter, 1982). The proposed expansion of the Park, which ultimately occurred in 1950, was accompanied by controversy and heated debate. Much of this debate was centered on management of the Jackson elk herd and was confounded by the supplemental feeding program on National Elk Refuge, which abuts the south boundary of the Park. The ultimate solution to allow public hunting within the expanded Park was unacceptable to the National Park Service. The Park Service made it clear from the outset that such a solution would be inconsistent with management policies for National Parks. Nevertheless, the Park Service and other groups believed an expanded Park was an extremely important addition to the National Park system. Thus, compromise provisions for dealing with the problem of managing the Jackson elk herd were included in Public Law 81–787 which established an enlarged Park.

Relevant sections of the Law follow:

> Sec. 6. (a) The Wyoming Game and Fish Commission and the National Park Service shall devise, from technical information and other pertinent data assembled or produced by necessary field studies of investigations conducted jointly by the technical and administrative personnel of the agencies involved, and recommend to the Secretary of the Interior and the Governor of Wyoming for their joint approval, a program to insure the permanent conservation of elk within the Grand Teton Park established by this act. Such program shall include the controlled reduction of elk in such park, by hunters licensed by the State of Wyoming and deputized as rangers by the Secretary of the Interior, when it is found necessary for the purpose of proper management and protection of the elk.
>
> (b) At least once a year between February 1 and April 1, the Wyoming Game and Fish Commission and the National Park Service shall submit to the Secretary of the Interior their joint recommendations for the management, protection and control of the elk for that year. The yearly plan recommended by the Wyoming Game and Fish Commission and the National Park Service shall become effective when approved by the Secretary of the Interior and the Governor of Wyoming, and thereupon the Secretary of the Interior and the Governor of Wyoming shall issue separately, but simultaneously, such appropriate orders and regulations, to be issued by the Secretary of the Interior and the Wyoming Game and Fish Commission, shall include provision for controlled and managed reduction by qualified and experienced hunters licensed by the State of Wyoming and deputized as rangers by the Secretary of the Interior, if and when a reduction in the number of elk by this method within the Grand Teton National Park established by this act is required as part of the approved plan for the year, provided that one elk only may be

killed by each such licensed and deputized ranger. Such orders and regulations of the Secretary of the Interior for controlled reduction shall apply only to the lands within the park which lies east of the Snake River which lie north of the present north boundaries of Grand Teton National Park, but shall not be applicable to lands within the Jackson Hole Wildlife Park . . .

Ever since its inception, there have been factions, either within the Park Service or outside, that have pursued eventual closure of the Park hunt (Murie, 1953; P. Wood, 1984). The most recent such action was the preferred alternative in the official plan for the Park released during the summer of 1985 in which an experimental reduction in the elk cull was proposed as a scientific experiment (Grand Teton National Park, 1985). The plan stated that the results of my research (presented in this book) should be reviewed before taking further action on this proposal.

Most of this chapter deals specifically with the Park hunt, since this is clearly the most direct action taken by the Park Service in managing the elk herd. There were other management issues addressed in Public Law 81–787, however, which affect the Jackson elk herd. For example, the Law provides for continuation of domestic livestock grazing in parts of the Park (Figures 8.1 and 8.2). The intent of the legislation is to eventually eliminate domestic livestock grazing in the Park; however, livestock driveways across the Park are to be provided as long as they are needed. Since 1950, the area grazed by cattle within the Park has been reduced 59% (Grand Teton National Park, 1985).

At the present time, there are eleven grazing permits totalling approximately 7500 AUMs (animal unit months) of grazing on 11 268.4 ha within the Park (Table 8.1). Most of these grazing permits will be phased out with the death of lease holders and their immediate descendents. Cattle grazing coincides with elk calving in the Uhl Hill and Two Ocean Lake areas (see Figures 4.20 and 8.2), thus the risk of transmitting brucellosis to cattle is particularly high. Cattle are placed on summer ranges within the Park in late May or early June. About 6300 cattle are trailed across the Park either by ranchers holding grazing permits for Park lands or to gain access to lands on the adjacent Bridger–Teton National Forest. Most trailing occurs during June and October.

When Grand Teton National Park was expanded in 1950, it included many private inholdings some of which will remain in private ownership, but most were, or are, under leases of varying duration. Much of the original private land is now owned by the Park Service, with the previous owners and their heirs having life estates. Only about 1480 ha or 1% of the Park is currently owned by

Figure 8.1 Cattle grazing south of Moran in Grand Teton National Park.

private individuals or by the State of Wyoming (Grand Teton National Park, 1985).

Public sentiments toward the Park hunt

Public reactions to the elk reduction program in Grand Teton National Park are mixed. The hunt is generally popular amongst hunters, and there is no shortage of hunters willing to participate in the Park hunt because hunter success is usually high and the area is relatively accessible. The hunt has been billed as an excellent example of interagency cooperation to ensure proper wildlife management (Thomas, Crowe & Kruckenberg, 1984).

Other individuals and groups strongly disfavor the program (Wood, 1984). This view was publicized in 1983 by CBS in a *60-minutes* television program entitled 'A Sporting Chance', which also criticized the elk hunt on the National Elk Refuge. Some criticisms of the Park hunt are from anti-hunting groups which view sport hunting anywhere as unacceptable (Amory, 1974). Others believe that the herd reduction program violates Park management philosophy (Ise, 1961).

Some public objection to the Park hunt comes from its visibility, particularly

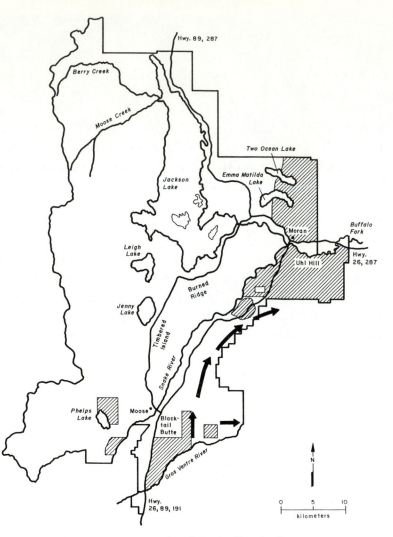

Figure 8.2 Grand Teton National Park. Hatched areas are livestock grazing areas. Heavy arrows are cattle trailing routes for access to National Forest lands.

in the Hayfields area near Mormon Row (Righter, 1982). Here elk migrating to the National Elk Refuge in late autumn cross open fields and sagebrush grasslands and are intercepted by hunters. Although few tourists are in the area at this time (Figure 8.3), hunters are often highly visible to the public and negative publicity results.

Evidence for the necessity of the Park hunt

Over the years, the approach which the Wyoming Game and Fish Department and the National Park Service have taken toward the Park elk reduction program has changed. Early on, the Park hunt was viewed primarily as a tool for ensuring adequate harvest of elk migrating from Yellowstone National Park to the National Elk Refuge (Cole, 1969). But during the late 1950s and early 1960s, the number of elk summering in Grand Teton National Park increased substantially (Martinka, 1969), so that now harvests in the Park are designed to control the number of elk that summer in Grand Teton National Park.

Part of the Department's justification for the Park hunt is to secure adequate harvest on the Jackson elk herd to keep the herd below the population limits established by the Wyoming Game and Fish Commission for feedgrounds in the valley. Indeed, particularly during mild autumns, harvests of elk may be inadequate to check population growth. The Park hunt becomes particularly important during these years of low harvest, as demonstrated by the highly significant inverse correlation between the proportion of total kill coming from the Park and total kill in the Jackson herd unit between 1961 and 1984 ($r = -0.525$, $n = 24$, $P = 0.008$).

Table 8.1. *Livestock grazing in Grand Teton National Park, 1984.*

Permitee	Cattle	Horses	AUMs**	Total area (ha)
Ida B. Chambers	235		520	185.0
Ernest H. Cockrell	150		532	1 672.1
Frank F. Galey		68	208	364.4
Jeannine P. Gill*	350		2100	2 169.2
Martha C. Hansen*	279		1673	1 712.6
Mary M. Mead*	290		1741	1 826.7
John W. Mettler, Jr		50	284	64.8
G. & M. Moulton	299		262	2 226.7
L. M. Rockefeller		25	111	160.3
Mr Rudd		60		259.1
Triangle X Ranch				607.3
Concessioner and GTNP		300		
Recreational use		50	108	20.2
Totals	1603	553	7539	11 268.4

*These three allotments overlap in the Uhl Hill area.
**AUM = animal unit month, i.e., 1 cow and calf on range for 1 month.

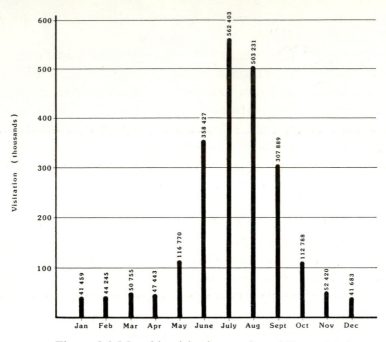

Figure 8.3 Monthly visitation to Grand Teton National Park in 1984.

Kill of elk within the Park is inversely correlated with estimates of the central valley elk population the following summer (Chapter 3). Likewise we find an inverse correlation between the combined per capita kill rate within the Park and Refuge and valley counts the following summer. These observations verify the effectiveness of the Park kill at reducing numbers of elk summering in Grand Teton National Park. The substantial increase in the proportion of bulls among elk summering in the Park, as a consequence of selective killing of antlerless elk during 1975–83, also indicates that the elk reduction program influences elk summering in the Park.

As shown in Chapters 2 and 5, early migrants are mostly elk from the Park; therefore, early season harvests within the Park and on the Refuge probably take mostly Park elk. A number of studies have shown strong philopatry among elk, particularly adult females (Craighead, Atwell & O'Gara, 1972; Shoesmith, 1979; Rudd, Ward & Irwin, 1983; Edge & Marcum, 1985), that is, elk have a strong tendency to return to the same summer range year after year. Therefore, the Park hunt may be necessary to ensure a harvest of elk summering within the Park. However, this possibility cannot be evaluated without more careful study of yearling dispersal, which could reduce the 'need' for the Park hunt. Indeed, many details of density dependence in elk are poorly understood.

For example, we do not know the mechanisms creating substantial variation in elk density within the Park.

Existing data on yearling dispersal may actually imply reinforcement for the 'need' for a herd reduction program in the Park (see Chapter 3). Although movement of yearlings out of the valley is density dependent, even during high-density years, there are more spike bulls that remain in central valley portions of Grand Teton National Park than can be accounted for by recruitment (Figure 3.15). Therefore, a fraction of the yearlings fail to migrate back to summer ranges on which they were born, but rather remain in the valley. This tendency is even greater during wet springs following low densities of elk on the National Elk Refuge.

Even though the Park kill is effective at culling elk from Grand Teton National Park, this does not imply that the Park hunt is necessary for satisfactory management of the Jackson elk herd. The most compelling argument for the 'necessity' of the Park hunt may be to ensure an adequate harvest to control the number of elk on winter feedgrounds and to restore historical distribution.

Evidence that Park hunt closure may be acceptable

Summer counts of elk in central valley portions of the Park have not changed substantially despite variation in harvest levels. Nevertheless, after an increase in antlerless harvests between 1975 and 1983, there was a significant decline in the cow-calf-yearling male segment of the herd while the mature bull segment of the herd increased in number.

The yield curve at Figure 3.33 suggests that closure of the Park hunt will result in an increase of 30–50% in the summer population, with carrying capacity at approximately 2000 elk in the central valley. We have documented density dependence at several levels, each contributing to the yield curve at Figure 3.33. These levels include density-dependent recruitment, calf survival and possibly yearling dispersal (Chapter 3).

Density-dependent recruitment may occur because of density-dependent fecundity or density-dependent calf survival as shown in Chapter 3. However, recruitment rates are relatively high for the Jackson elk herd, and the slope of the density dependence is shallow. If one were to assume linear density dependence, the population would continue to grow to many times its present size before reaching limits set by density-dependent recruitment. However, such a linearity assumption is almost certainly incorrect. In all other vertebrate populations for which sufficient data are available, a concave density dependence relation exists (Fowler, 1981; Boyce, 1984), that is, as density increases above levels observed thus far, recruitment rates may decline much more rapidly. This same discussion applies exactly for density-dependent survival of young calves.

A more responsive mechanism may be density-dependent dispersal. Free-ranging elk which were not provided with supplemental food during winter dispersed from high density winter ranges in the Firehole area of Yellowstone National Park (Cole, 1983) and from portions of the northern Yellowstone winter range (Houston, 1979, 1982). However, such a population-regulating mechanism seems irrelevant to the Grand Teton National Park population, since virtually all of these elk winter on the National Elk Refuge where supplemental food is provided. Winter dispersal is unlikely under these conditions.

Density-dependent dispersal of yearlings may be an important population-regulating mechanism for the Park herd segment. There is limited evidence for density-dependent dispersal for the small National Elk Refuge summer herd, or at least a failure of yearlings to migrate following winters of low density on the Refuge. However, similar evidence could not be demonstrated over the entire herd (see Chapter 3). Clutton-Brock, Major & Guinness (1985) failed to find any evidence for density-dependent dispersal of yearlings as a population-regulating mechanism for red deer on the Isle of Rhum. Nevertheless, density-dependent yearling dispersal certainly may be important, and failure to find evidence for it in two instances does not preclude its possible importance at higher densities in the Park.

Risks of terminating the Park hunt

A proper survey of public attitudes towards the Park hunt has not been conducted, but I submit that a majority of Park visitors would be opposed to allowing hunting within Grand Teton National Park. Certainly, there is strong opposition to the hunt from some interest groups while other groups support the hunt. A very real risk to an experimental closure of the elk reduction program is that public sentiment could make it an irreversible management decision. Public sentiment to reopening the hunt after closure may be extremely strong, despite possible negative consequences having resulted from a closure.

Establishing criteria for justifying restitution of the Park hunt may be difficult. To some of the public, if there is still green grass and elk in the Park in summer, the consequences of high elk numbers will not have been great enough to justify the Park hunt. Procedures for estimating elk numbers and range condition may have their weaknesses, but the critical issue is deciding what levels of elk population size or range deterioration are truly adequate to reinstitute elk control. Extremely high elk numbers may be necessary to cause damage to summer range, and once such elk densities have been attained, it may require a very long time for the range to recover. But it is unlikely that such levels will be more extreme than occur ubiquitously in situations of heavy cattle use which occur throughout much of the western United States. Establishing criteria for unacceptable levels of damage to plant communities in the Park must be subjective. Consequences to other wildlife may be mostly detrimental, although various predator populations will probably be enhanced.

Even with liberal seasons and regulations from 1974 to 1983, the average number of elk killed in the Park was only 680 elk, with 37% hunter success. A temporary discontinuation of the elk reduction program may result in a substantial increase in the size of the Park herd segment. Securing an adequate harvest of this enlarged herd may be difficult after the reduction program is reinstituted. This could be readily resolved by killing elk west of the Snake River, but this would require an act of Congress to override the specifications in Public Law 81–787.

Similar to this argument is that without the Park herd reduction program, in many years it may be difficult to secure an adequate harvest of elk from the herd unit to maintain control over feedground numbers. Statistical evidence for this is presented in Chapter 7.

Are there alternatives to a cull?

Various methods of birth control for elk, such as chemosterilants, interuterine devices, or ovariectomies have been studied (Greer, Hawkins & Cutlin, 1968), but would be impractical for controlling the number of elk that summer in the Park. The National Elk Refuge, during winter, is the only feasible place to handle large numbers of elk, but once elk are on the Refuge, those that summered in the Park could not be distinguished from other elk that should not be sterilized.

Reduced feeding of elk may accomplish similar results much more efficiently. Unfortunately, we are still faced with the difficulty of separating elk that summer in Grand Teton National Park from non-target elk. Reducing the feeding program will reduce recruitment for the entire herd and consequently reduce the allowable harvest. At the present time, public sentiments are strongly against a reduction in the feeding program.

Another alternative that was suggested to me includes extensive fencing to reduce mingling between various herd segments (E. Wampler, Concerned Citizens for the Elk, pers. comm.). This would be very expensive, and would not alleviate the need to cull Park elk to maintain a balance of elk from various herd segments on the feedgrounds.

Strategies to improve the Park hunt

Some objections to the Park hunt could be circumvented if public visibility were reduced and if the quality of the hunt were improved to ensure a quality hunting experience for participants as well as other Park visitors.

More effective harvests of elk in the Park could be secured if hunting were allowed west of the Snake River. Many elk often evade hunters by staying west of the River during the hunting season, crossing the river and moving onto the National Elk Refuge at night when hunting is not allowed. Hunting on the west side of the Snake River would require special congressional action, however, because the original legislation specifies that in valley areas of the Park hunting will only be permitted east of the Snake River. Also, hunting on the west side of the Snake River would impinge on areas of the Park that are more heavily used by traditional Park visitors, although visitation is low during hunting season (Figure 8.3).

Hunting west of the Snake River on a strictly controlled basis would enable harvesting of Grand Teton National Park elk and would provide a higher quality hunting experience than hunting the Hayfields because the elk would be dispersed on their summer/fall range in habitat with substantial escape cover. By designing seasons to precede the migration of Teton Wilderness and Yellowstone National Park elk into Grand Teton National Park, the Grand Teton National Park herd segment would be targeted. A well-designed hunt west of the Snake River would reduce the need for the Hayfields hunt.

Hayfields

One possible strategy for improving the Park hunt and reducing public visibility is to close the Hayfields area to hunting. This is an open area in the Park immediately north of the National Elk Refuge, where culling occurs during the migration to the Refuge. The argument against this approach is that a substantial fraction of the Park hunt is often secured in the Hayfields area. Between 1974 and 1984, hunter kill from the Hayfields ranged from 7% to 57% of total Park harvests, and averaged 27%.

When asked about the low quality of the Hayfields hunting experience, Park Service and Wyoming Game and Fish Department employees have told me that the Park hunt is not a sport hunt but rather it is a herd reduction program. The Park Service has no interest in fostering high quality hunting experiences, but it is the official policy of the Wyoming Game and Fish Department to give priority to the quality of the outdoor recreation experience (Wyoming Game and Fish Department, 1983). With growing nationwide sentiment against hunting, I submit that it is important to ensure high quality hunting experiences whenever possible to reduce public opposition. Semantic differences between sport hunting and elk herd reduction program seem irrelevant. Objectionable modes of hunting are likely to evoke strong public reaction.

In recent years, the visibility of the Hayfields hunt has been reduced by closing the Mormon Row road during hunting season, and by making it illegal to discharge a firearm within 0.4 km of Highway 26/191. It is unlikely that visitors to Grand Teton National Park during hunting season will witness an elk being killed, although they will certainly see hunters. Blaze orange garments which hunters are required to wear make them highly visible.

Berry Creek

Perhaps the highest quality hunt that one could obtain in Grand Teton National Park would be a wilderness hunt in the large Berry Creek–Moose Basin area (19 845 ha) in the northwestern portion of the Park. Hunting is expressly permitted there by the enabling legislation, and the area has been designated hunt area 72 by the Wyoming Game and Fish Department. However, in the past when hunting was allowed there, harvests were low and the National Park Service did not feel that harvests were large enough to justify desecrating the area with sport hunters. Access to the roadless area is not easy, and is best obtained with a boat or with horses. Present Park regulations do not allow travel off of designated trails with horses, do not allow grazing, and only allow camping with horses at two designated sites in the Berry Creek area. When opened for hunting during eight seasons between 1950 and 1967, an average of only 15 (range 4 to 29) elk were killed annually and interestingly 36% of these were taken by employees of the Wyoming Game and Fish Department or other members of their hunting party (Cole, 1969).

Estimates of the number of elk in the Berry Creek–Moose Basin area vary from 180 in 1958–59 (Guse, cited by Barmore, 1984) to 300 in 1964 (Cole, 1969: 16). Numerous reconnaissance trips into the area and counts and estimates of elk were summarized by Barmore (1984), who concluded that densities of elk in the area were not high, and that there is no evidence of over-utilization of range. Migration of elk from the Berry Creek area can be either along the west shore of Jackson Lake or across the Snake River north of the Lake and then south along its east side (Houston, 1968; Yorgason & Cole, 1967). Therefore, track counts on the Burned Ridge transect may include Berry Creek elk some years, whereas in other years these elk may cross the Snake River or Pacific Creek transects (see Chapter 2).

Given the insignificant harvests from the Berry Creek area, it seems impossible to justify conducting a herd reduction program there. Certainly hunts in the Berry Creek area cannot fulfil the requirement of Public Law 81–787 that it be 'necessary for the purpose of proper management and protection of the elk'. The intent of Public Law 81–787 is clearly not to sustain high quality wilderness hunts, but rather to ensure culling the elk herd when necessary. Harvests from the Berry Creek area contribute little to this cull.

Philosophy of Park management and the preservation of biotic diversity

Grand Teton National Park is clearly not a complete ecosystem; human intervention is present in the form of cattle grazing allotments, an important natural predator, the wolf, has been exterminated from the area, and nearly all elk that summer in the Park winter on the National Elk Refuge where they are usually fed. Most elk probably have always migrated beyond the area encompassed by present Park boundaries during winter (Cole, 1969).

One of the objectives for maintaining conservation areas is to preserve biotic diversity (Myers, 1979). Indeed, the Park gives management priority to the few threatened or endangered species that occur within the Park (Grand Teton National Park, 1985). Grizzly bears are uncommon in the Park, most occur further north in Yellowstone National Park. Trumpeter Swan (*Olor buccinator*), Peregrine Falcon (*Falco peregrinus*), and Bald Eagle (*Haliaéetus leucocéphalus*) all occur in the Park. None of these species is likely to be influenced by the elk management program, therefore, loss of genetic diversity is not likely an issue in determining the need for or magnitude of any elk reduction program in the Park.

Rather, the more important role for Grand Teton National Park is as an ecological baseline area in which manipulation by man is minimal to ensure existence of an area for comparison with surrounding areas. Sinclair (1983) suggests that even if it is not possible to encompass a complete ecosystem within Park boundaries, and even if human intervention has altered part of the system, it is desirable to have conservation areas in which human intervention is minimal, and where the remainder of the system is self-regulating. However, it is possible that non-intervention with the elk population in the Park might alter the system from its pre-history state even more than continued culling of Park elk (Chase, 1986), but this seems highly unlikely (Cole, 1983).

Unfortunately, records of early populations of elk in Jackson Hole are inconsistent and even contradictory. Fossil and archaeological evidence shows that elk have been in the region for millennia (Bryant & Maser, 1982), and have been used by native Americans since at least Folsom times (Frison & Bradley, 1981). Wapiti remains have been dated at 4000 years BP from the Dead Indian Creek archaeological site in the Absoroka Mountains northeast of Jackson Hole (Scott & Wilson, 1984). However, the relative abundance of elk prior to

1912 is unknown. For example, old letters and diaries of Beaver Dick, an early trapper in the valley, imply that elk numbers were apparently quite low (Murie & Murie, 1966). But again, early records seem inconsistent (for reviews see Anderson, 1958; Cole, 1969).

Just prior to the turn of the century, a large number of elk wintered in Jackson Hole (Sheldon, 1927; Preble, 1911). Yet, these large numbers of elk may have been partly a consequence of man-induced perturbations, for example, predator control and habitat alterations caused by burning and grazing. A majority of the aspen stands in Jackson Hole originated from extensive fires during the 1870s and 1880s (Gruell, 1979; Hart, 1986). Peak forage production from aspen stands occurs within a few years after fire (DeByle & Winokur, 1985). Although by itself aspen may not have had a major consequence to elk population dynamics, combined vegetation changes caused by man may have been important.

The salient point is that we do not know how many elk occupied the Park prior to human disturbance, and consequently we cannot know the proper culling rate for elk in the Park. We do not know if predation by wolves maintained elk at much lower densities than now. Wolves have long existed in Jasper National Park in the Canadian Rocky Mountains, but densities of elk on winter ranges there still reach such high levels that culling was thought 'necessary' to reduce competition with bighorn sheep on winter range (see Flook, 1970; Stelfox, 1976). Likewise, Carbyn (1983) found that wolf predation on elk in Riding Mountain National Park in Manitoba was inadequate to check the growth of the elk herd.

Some analogies may exist with management of elk in northern Yellowstone National Park. Houston's (1982) justification for non-interference management of the northern Yellowstone elk herd is an attempt to maintain the ecosystem as near to pristine conditions as possible. Yet, he recognizes that wolves have been eliminated from the system, and without wolves, the system cannot be exactly like it was in 1850.

Pengelly (1963) stated that the northern Yellowstone range could support no more than 5000 elk. Despite the fact that 5000 was the Park Service's estimate of carrying capacity and was based upon the opinion of a group of biologists at the time, Sinclair (1983) suggests that the actual basis for such figures is really a mystery. Houston's (1982) research has shown that the range can certainly sustain much larger numbers of elk. Likewise, without adequate evidence, Beetle (1974a, 1979) insists that there are too many elk in Jackson Hole. How do we know that the Wyoming Game and Fish Department herd objective of 11 000 elk isn't just right? Or twice that number? Tragically, we do not know.

In the Plan for Grand Teton National Park released in summer 1985, one of several proposed management actions is to experimentally reduce the number of elk culled in the herd reduction program. Such a program will obtain more data points for the right-hand side of the yield curve presented in Figure 3.33. As emphasized above, and as pointed out in the Plan (Grand Teton National Park, 1985), we cannot be certain of the consequences. Yet, the only way to evaluate the efficacy of such management programs is through experimental manipulation (Houston, 1982).

The hunt in Grand Teton National Park creates a real dilemma. Although it would be interesting to experiment with a closure of the Park hunt, the consequences will almost certainly include an increase in the number of elk summering in the Grand Teton National Park. This will necessarily result in a decline in the number of elk summering elsewhere in the herd unit because the number of elk on the herd unit is constrained by the quota for the feedgrounds. Thereby, the Park Service is essentially confronted with trading elk in Grand Teton National Park for elk in southern Yellowstone National Park. Park administration aside, there is also value to maintaining elk populations in the Teton Wilderness and other parts of Bridger–Teton National Forest. The Park hunt affords a vehicle for maintaining better distribution of elk throughout the Jackson herd unit.

Summary

1. When Congress expanded Grand Teton National Park in 1950, it included a provision that a cull of elk would occur if deemed necessary by the Park Service and the Wyoming Game and Fish Department. The 'necessity' of the hunt has been debated since its inception.

2. Cattle are grazed during summer within Grand Teton National Park on an elk calving area. Risk of brucellosis transmission from elk to cattle is high in this area, but could be substantially reduced by postponing release of cattle into the area until 30 June of each year.

3. Hunting within the Park has been highly visible and severely criticized in the past. As a consequence, more stringent regulations on hunting have reduced objectionable killing practices.

4. Increased numbers of elk in Grand Teton National Park has reduced the numbers of elk that can be hunted on the National Forest because quotas for total population size are set for winter feedground numbers.

5. Natural regulation of the Jackson elk herd has been altered because of the winter feeding program. Currently, it is the quota on feedground numbers that imposes population limitation rather than

natural factors such as winter habitat. Consequently, reducing the Park hunt will result in increased numbers of elk summering within Grand Teton National Park, whilst continued culls outside the Park will reduce the number of elk in southern Yellowstone National Park and on the National Forest. This creates a dilemma for the Park Service whereby tradeoffs occur for the numbers of elk maintained in Grand Teton National Park versus Yellowstone National Park.

Chapter 9.

Public interests in elk management

Management of wildlife by government agencies ultimately reflects various values of the wildlife resource to the public. Ideally, management directives reflect a balance of interests in the resource. For the Jackson elk herd, public interest in hunting, viewing, conservation, and photography favor protection and enhancement of elk and their habitats, whilst other public demands for mining, commercial, residential and recreational development, livestock grazing, and timbering may conflict with elk.

Here I summarize some of the public interests in elk, beginning with consumptive uses such as hunting, guiding and outfitting, and various activities that compete for elk habitats such as logging, mining, and land development. Concluding remarks are offered on the role of elk in the Greater Yellowstone Ecosystem.

Economic interests

The Jackson elk herd is unquestionably important to the region's economy, either directly or indirectly. People like to see elk, to hunt them for trophies and meat, to photograph them, or just to know that they are there. Hunting and observing wildlife attracts millions of dollars to the valley each year. At the same time, other interests of economic importance may indirectly compete with elk by altering habitats or displacing elk from ranges. These principally entail the timber industry, oil and gas exploration, and recreational use of elk summer and winter ranges.

Elk hunting

For over a century, recreational elk hunting has been popular in Jackson Hole, and the area is renowned for pack trips into remote wilderness areas in pursuit of the wily wapiti. Theodore Roosevelt considered the area within an 8 km radius of Two Ocean Pass to be the finest big game area in the world (Taylor, Bradley & Martin, 1982).

Estimating economic value of elk hunting requires numerous considerations, including value of meat, hide and antlers, and also expenditures by hunters for equipment, lodging, food and travel. Strictly based upon average expenditures by elk hunters, Phillips & Ferguson (1977) estimated that the total value of elk hunting in the state of Wyoming in 1970 was $10.5 million. Resident elk hunters spent on average $147 and nonresident elk hunters spent $693 during elk hunting trips. Much of the greater expenditures by nonresident hunters was due to nonresidents being more likely to employ an outfitter or guide. In 1980, nonresident hunters using a guide in Jackson Hole spent an average of $1723 during their hunting trip (Taylor, Bradley & Martin, 1982).

Since the Phillips & Ferguson (1977) study, methods for estimating the value of wildlife and hunting have been refined. Hansen (1977) used the contingent valuation method to value elk hunting in the Rocky Mountains at $36.37 per day adjusted to a 1982 base. Sorg & Loomis (1985) advise that this estimate should be adjusted upward by 20% to account for a biased sampling procedure employed in the Hansen study, yielding a value of $43.64 per visitor day.

During 1982, elk hunters spent 69 261 recreation days hunting elk in Jackson Hole. Based upon the Sorg & Loomis (1985) estimated valuation, this suggests total expenditures of $3 022 550. However, the area studied by Hansen (1977) in Utah did not require nonresident hunters to have a guide or outfitter. Because all nonresident hunters in wilderness areas of Jackson Hole are required by law to employ a guide, this value may underestimate the economic value of elk hunting to the area.

Most elk hunters in Wyoming valued hunting for meat higher than other reasons for hunting. Only slightly more than 1% gave top priority to obtaining a trophy elk (Phillips & Ferguson, 1977). In view of these priorities, it is difficult to justify decreasing the amount of elk hunting opportunities in order to manage for more trophy elk (Phillips & Ferguson, 1977). This survey also indicated that most elk hunters prefer bulls-only restrictions on hunting to limited quota permit constraints.

Although few hunters rank trophy hunting as their principal objective in elk hunting, most hunters prefer to shoot mature bulls (Anderson, 1958). Jackson Hole has produced more trophy-class elk than any other area in North America. Currently at least 14 elk from the Jackson herd are listed among trophy elk registered by the Boone and Crockett Club (Table 9.1). Note that among registered trophies, more trophy-class elk have been killed in Jackson Hole since 1960 than during previous years. This implies that the quality of trophy animals killed has not deteriorated in recent years. The largest bull elk to be killed in Jackson Hole was killed in 1972, ranking 12th in North America.

Outfitters and guides

Hunting success for elk is lower than for any other species of deer in North America, although the probability of killing an elk is usually higher when accompanied by an experienced guide (Whitehead, 1982). Guided wilderness pack trips for elk hunting are a long-standing tradition in Jackson Hole, and the outfitting and guiding business sustains well over 2 million dollars for the local economy (Taylor, Bradley & Martin, 1982).

The local guiding and outfitting industry is subsidized by Wyoming state law which requires that nonresident hunters be accompanied by a licensed resident or professional guide and outfitter to hunt in any designated wilderness area. This partly accounts for the fact that 94% ($n = 1391$) of all hunters using outfitter and guide services in Wyoming in 1980 were nonresidents (Taylor, Bradley & Martin, 1982). In 1980, hunters using outfitter or guide services paid an average of $1218 on services and a total of $1723 for the entire hunting trip. Total expenditures by guided hunters in Jackson Hole in 1980 equalled $2400000 (Taylor, Bradley & Martin, 1982). The actual impact on the local

Table 9.1. *Record-book trophy elk killed in Jackson Hole.*

Year	Rank	score
1912	61	$388\frac{7}{8}$
1936	46	$391\frac{6}{8}$
1947	17	$400\frac{2}{8}$
1950	215	$375\frac{3}{8}$
1954	220	$375\frac{2}{8}$
1956	101	$384\frac{3}{8}$
1957	215	$375\frac{3}{8}$
1962	184	$377\frac{3}{8}$
1963	66	$387\frac{7}{8}$
1964	163	379
1966	184*	$376\frac{5}{8}$
1968	195	$376\frac{6}{8}$
1970	117	$382\frac{4}{8}$
1972	12	$400\frac{2}{8}$
?	193	$376\frac{7}{8}$

Scoring measurements and names of owners for each of these trophies may be found in Nesbitt & Wright (1981).
*Listed by Nesbitt & Parker (1977) but not Nesbitt & Wright (1981).

economy is even greater because monies are respent among local businesses and households. A multiplier derived by Premer *et al.* (1979) yields an estimated total contribution to the local economy generated from outfitted hunting in Jackson Hole during 1980 to be $4.2 million.

Income and business generated by the outfitter and guide industry is of particular importance in Jackson Hole because of the seasonality of employment and gross sales in the valley. Revenues generated during September–November hunting help dampen the strong seasonal nature of annual sales (Figure 9.1). Most guides and outfitters continue to work in summer, often offering float trips, pack trips or dude ranch operations.

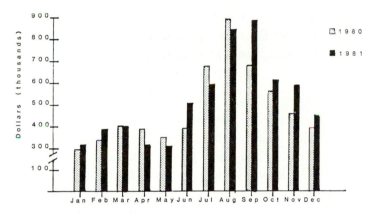

Figure 9.1 Monthly distribution of sales and use tax receipts in Teton County, Wyoming, 1980–81. Hunting during October and November reduces the seasonal amplitude in sales from that suggested by Park visitation (see Figure 8.3). Wyoming sales and use tax data were obtained from the University of Wyoming Institute for Policy Research, Data Retrieval System.

Because outfitters and guides contribute significantly to the economy of Jackson Hole, and because they have monetary incentive for management of the Jackson elk herd, they play an important role in the politics of elk management. In 1985, the Jackson Hole Outfitters and Guides Association submitted the following recommendations to the Wyoming Game and Fish Department for use in planning. Another smaller organization, the Concerned Citizens for the Elk, did likewise.

The first six recommendations are shared by both organizations, with an additional four offered by the Concerned Citizens for the Elk. Following each recommendation, I evaluate the recommendation within the context of the data available on the Jackson elk herd. Some of these recommendations are supported in my final management recommendations for the Jackson elk herd in Chapter 10. The recommendations are:

(1)   *To increase the herd unit objectives for the Jackson herd from 11 200 to 12 500. This entails approximately a 12% increase, which will hopefully not come from elk summering in Grand Teton National Park. Because the Outfitters and Guides Association believes the number of summering elk in the Park to have increased, to sustain similar numbers of elk in non-Park areas it is argued that increased total numbers of elk must be maintained on feed.*

The motivation for this recommendation is probably to increase hunter success or number of hunters. Increasing the elk population will increase the cost of winter feeding, and may risk causing even further declines in the Gros Ventre bighorn sheep population, for example, on Sheep Mountain. Evidence from Figure 3.33 suggests that the Jackson elk herd has been maintained at levels near to maximum sustained yield, and increasing herd size may therefore be counter-productive.

(2)   *To offset costs of feeding elk in Jackson Hole, it is recommended that the state institute a $10 elk feed stamp, required of all individuals hunting in areas of northwestern Wyoming where supplemental feeding is used to sustain larger elk populations. It is also suggested that a higher fee be instituted for licenses to hunt in Grand Teton National Park and on the National Elk Refuge.*

This suggestion is perfectly coincident with policy stated in the Wyoming Game and Fish Department's (1983) five-year Strategic Plan, that is, 'to develop and implement methods whereby the users of high cost production projects pay their fair share of the management costs'. Strongest resistance will probably come from resident hunters for whom this will constitute a 40% surcharge on the existing license fee.

(3)   *To establish an additional elk feedground in the valley of the Buffalo Fork River, east of Moran, Wyoming. It is suggested to build this herd segment to approximately 1000 animals.*

As reviewed in Chapter 5, I find it appealing to potentially separate part of the migratory Yellowstone–Teton Wilderness herd segment from the Grand Teton National Park herd segment, because it would improve control over different parts of the population. Unfortunately, however, the recommended location for another feedground is unacceptable because (a) either competition will occur with 150 moose that winter in the area or moose will need to be fed as well, (b) there would be inevitable impacts to elk calving habitats in the Buffalo Fork valley, (c) geography of the valley is such that heavy trailing and consequent soil erosion is inevitable, (d) there is risk of transmitting brucellosis between elk and cattle that winter in the valley, and (e) the area is used by grizzly bears.

(4)   *To shorten hunting seasons and restrict kill to bulls only in some areas where the herd appears to have been depleted, for example, east of Chicken Ridge in the Teton Wilderness in hunt area 70.*

The migration across track transects east of Togwotee Lodge was the greatest in the valley prior to 1960 (see Chapter 2). Sustained heavy harvests on this herd segment have reduced it to the smallest component of the elk migration. The Togwotee migration is earlier than migration further west in the herd unit and therefore elk are usually crossing Highway 26/287 while the season is open. In most years, elk migrating later, further west, cross the Highway in November and therefore are not subject to hunting if the season closes as it often has on 31 October.

Unfortunately, restricting harvests on area 70 may be inadequate to rebuild this herd segment, because the elk migrate through several other hunt areas en route to wintering grounds. Harvests in area 70 and along migration routes should be restricted for at least ten years to allow for a response because of the difficulty in constraining harvests on these elk. Reduced road access on selected National Forest lands may also speed recovery for this herd segment.

(5)   *To reopen hunting seasons in the Berry Creek–Moose Basin area of north-western Grand Teton National Park.*

This issue is discussed in Chapter 8, and I conclude that the Berry Creek unit 72 hunt is unacceptable and inconsistent with the motivation for the Park hunts. Berry Creek should not be reopened to sport hunting, because harvests there do not contribute enough to the objective of conserving the elk herd to justify adverse impacts on Park resources.

(6)   *To reduce hunter access and disturbance to elk on areas of Bridger–Teton National Forest, it is suggested that roads be closed after 1 October each year.*

The apparent motivation behind this recommendation is to reduce harvest by hunters who do not employ an outfitter or guide and to increase the demand for outfitter services. Road closure is an extremely useful tool to ensure that elk are not disturbed by human activity along roads and therefore are able to use a greater fraction of available habitat. Gruell & Roby (1976) showed that it was human disturbance and not the mere existence of the road that precluded elk use. A blanket recommendation that roads be closed on 1 October is ill-advised because disturbances along roads will still occur during the entire summer and early autumn period.

In addition to these recommendations, the Concerned Citizens for the Elk suggest:

(7)   *To close all hunting seasons on 31 October of each year.*

Such a constraint on the timing of hunting seasons is undesirable because it reduces harvest options. To obtain a cull on elk migrating from Yellowstone National Park, hunting after 31 October is usually necessary, since peak migration is usually after that date. The appropriate time to hunt the migratory Yellowstone–Teton Wilderness herd is during migration in November when Yellowstone elk are more likely to be killed. Perpetual hunting during September and October is more likely to selectively remove elk that summer in the Teton Wilderness.

(8)   *To transplant surplus elk from the National Elk Refuge to state feedgrounds. They note that volunteer help and financial resources will be available for these transplant operations from the Jackson Hole Outfitters and Guides Association, the Concerned Citizens for the Elk, and the Rocky Mountain Conservation Fund.*

Even though volunteer help for this effort may be available, thereby reducing costs for a very expensive operation, there are several other reasons why such a program is objectionable: (a) most transplanted elk return to the area where they were originally trapped (Allred, 1950; Janson, 1966), (b) transplanting stresses the elk, and (c) there is no way to ensure that calves selected for transplanting are from target herd segments; therefore transplanting would not serve to reduce specific herd segments.

Many elk were transplanted from the National Elk Refuge in the past as summarized by Robbins, Redfearn & Stone (1982) and Allred (1950). Previous experience indicates that transplanting adult elk will almost certainly be unsuccessful because adults tend to return to their native range, sometimes exceeding 200 km (Janson, 1966). Therefore, transplanting Refuge elk would have to be confined to calves, but even calves may sometimes return to their site of capture (T. Toman, Wyoming Game and Fish Department, pers. comm.). A further drawback is that transplanted calves may suffer high mortality, particularly during tough winters.

Because of the high incidence of brucellosis in elk on the National Elk Refuge, transplants must remain within northwestern Wyoming, or be tested and certified brucellosis free. But most suitable elk habitat is now fully stocked, and transplanted animals are superfluous. A high proportion of transplanted calves are killed by hunters very near to the release site (Janson, 1966), thus it would appear that a transplant program may essentially entail 'put-and-take' management for elk. Most state feedgrounds in northwestern Wyoming are already near capacity, and transplanting elk to state feedgrounds would seem to serve no constructive function.

Transplanting elk is an extremely expensive and ineffective means of controlling the Refuge population. So long as hunter harvest remains a viable option for herd management, transplanting elk is unjustified.

(9) *To reinstitute tagging studies of elk at the National Elk Refuge.*

Tag return data can provide useful information on movement in and out of the herd unit. Over 11 000 elk have been tagged in Jackson Hole, but the existence of a large volume of such data does not diminish the value of future tagging studies.

(10) *To begin supplemental feeding operations earlier, to continue feeding longer, and to increase the daily per capita rations to 4.5 kg. It is also suggested that creep feeders be used to ensure supplemental feed for calves.*

Weight loss by cow elk can result in low calf survival (Thorne, Dean & Hepworth, 1976), but there is no evidence that feeding rations are currently low or that calves need creep feeders to obtain adequate feed. Survival rates for elk on the Refuge and Gros Ventre feedgrounds are high, and recruitment rates are no longer reduced by low feed rations, although they apparently were prior to 1960. Increasing the amount fed will only increase the already considerable cost of feeding with virtually no benefits. Many variables contribute to an optimal feeding ration, and flexibility in feeding should be maintained rather than prescribing a minimum ration. Current feeding operations at the National Elk Refuge with pelleted alfalfa as detailed by Smith & Robbins (1984) do not result in notable competition for calf elk, and therefore I see no justification for installing creep feeders.

In addition to these suggested recommendations, the Jackson Hole Outfitters and Guides Association acknowledges concern about competition between elk and bighorn sheep on sheep winter ranges. Increased numbers of elk may ultimately increase this competition even more, thereby reducing overwintering bighorn sheep populations in the Gros Ventre and near the National Elk Refuge.

Other interests

Consumptive use of the elk resource by hunters, outfitters and guides is clearly of major importance in Jackson Hole. However, most visitors to Jackson Hole that enjoy elk are nonconsumptive users, and their economic contribution outstrips that of hunters. In addition, economic interests in oil and gas mining, recreational development, housing development, and logging may influence elk by altering their habitat.

Economic contribution of Grand Teton National Park

Approximately two-thirds of the income, sales, employment and population in Teton County are attributable to Grand Teton National Park, and would be absent without the Park (Merrifield, 1983). The direct net impact of the Park on sales in Teton County amounted to $44.6 million in 1978 (Merrifield & Gerking, 1982). Discounting other possible activities such as forestry, grazing and real estate development which would occur if the Park were not present, it still contributed $23.1 million to the county in 1978 (Merrifield & Gerking, 1982).

Elk are only part of the attraction of Jackson Hole, and the importance of elk to reinforcing visitation to the valley is unknown. Yet, wildlife and scenery are clearly the chief attractions (Gladney, 1985; Taylor, Bradley & Martin, 1982). Total visitors to the Park since 1970 have ranged from approximately 2 million to over 4 million persons (Figure 9.2), and Grand Teton and Yellowstone National Parks are two of the ten most popular vacation destinations in the

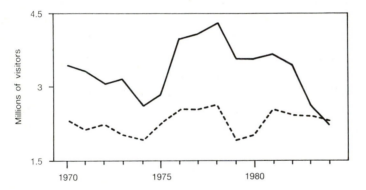

Figure 9.2 Visitation to Grand Teton National Park (solid line) and Yellowstone National Park (dashed line), 1970–84 (adapted from Gladney, 1985). A new estimation procedure was initiated in 1983, therefore estimates for 1983 and 1984 are not directly comparable to earlier values.

United States (Merrifield & Gerking, 1982). This visitation is somewhat misleading, however, because many visitors to Jackson Hole are driving through. In 1978, 50% of the summer visitors to Jackson Hole had Yellowstone National Park as their primary destination but Grand Teton National Park was the primary destination for only 26.4% (Fletcher *et al.*, 1979). Nevertheless, tourist-related expenditures accounted for between 59% (Merrifield & Gerking, 1982) and 70% (Blevins & Bryson, 1980) of Teton County's economy.

Commercial, residential and recreational development

After an extensive survey of human developments in Wyoming, Rippe & Rayburn (1981) concluded that community developments, particularly subdivisions, are the greatest threat to wildlife habitats, not only in Teton County, but in all of Wyoming. This is partly because many communities are in riparian habitats, which are often critical seasonal habitats for wildlife. Jackson Hole is no exception, with developments in the town of Jackson, on the Gros Ventre Buttes, and in South Park occupying some of the most productive land in the valley that was once prime winter range for elk.

Land in the Jackson herd unit is mostly under federal jurisdiction, and less than 3% is privately owned (Figures 1.5 and 9.7). Despite their relatively small area, however, some private lands are particularly important wildlife habitats. For example, 48% of the 350 km² of riparian habitat in Jackson Hole is privately owned (Wells, 1979). In fact, 66% of all private land in Jackson Hole is riparian, which is particularly important to wildlife. Because elk have often conflicted with agricultural interests, relatively few elk now occur on private lands within Jackson Hole.

Existing regulations governing the management of the Jackson elk herd were mostly established in the first half of the 20th century when the economy of Jackson Hole was dominated by agricultural interests, primarily cattle ranching. While ranching is still important in the valley, in recent years the population has grown substantially (Figure 9.3), and the economic base has

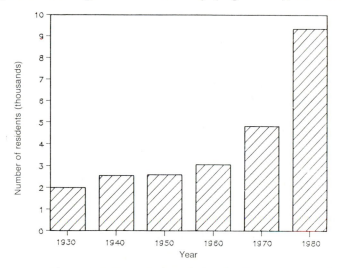

Figure 9.3 Human population in Teton County, Wyoming, from the National Census, 1930–80.

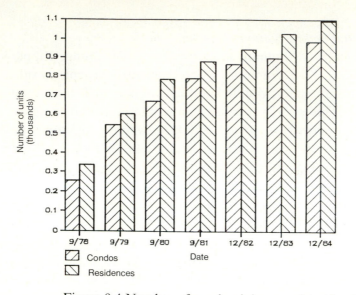

Figure 9.4 Number of condominiums and residential dwellings in Teton County, Wyoming, September 1976–December 1984.

shifted primarily to tourism. Land values have skyrocketed, condominiums and residential dwellings have quadrupled over the past eight years (Figure 9.4), and the acreage in housing subdivisions has more than tripled since 1976 (Figure 9.5). During 1967–77, slightly more than 1% of the crucial elk winter habitat in Jackson Hole was disturbed by land development (Rippe & Rayburn, 1981).

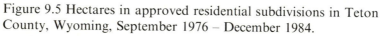

Figure 9.5 Hectares in approved residential subdivisions in Teton County, Wyoming, September 1976 – December 1984.

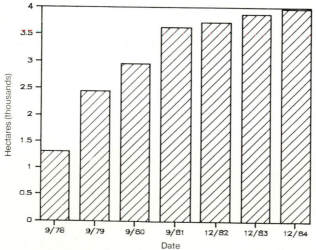

Although economic priorities still prevail in many development considerations in Jackson Hole, greater concern now exists for ensuring high quality development to maintain the scenic beauty of the valley. Nevertheless, land development, particularly residential subdivisions, has the greatest potential for destroying wildlife habitat in Wyoming, and Jackson Hole is no exception (Rippe & Rayburn, 1981).

Some regulation of private land uses in Jackson Hole is accomplished through zoning and permit requirements of the city and Teton County (Teton County Board of Commissioners, 1980), or in rare instances where adequately justified, by government condemnation of private property. In other instances, landowners may agree to sell their land to government agencies, to trade parcels of land, to sell development rights or to grant conservation easements. Such agreements have played a major role in the growth of the National Elk Refuge, and securing the integrity of Grand Teton National Park and Bridger–Teton National Forest.

In addition to private and federal lands, the Jackson elk herd unit encompasses about 1700 ha of state lands, which are administered by the State Land Board. Although the State Land Board could sell state lands or permit uses and development that may be detrimental to elk, there appear to be few specific elk management issues within the Jackson herd unit involving state lands. One exception is trespass grazing of livestock on state-owned land surrounding the Patrol Cabin feedground in the Gros Ventre River valley (see Chapter 6).

Many threats of future development in Teton County are also on elk critical winter range. For example, the Bureau of Land Management and the US Fish and Wildlife Service have arranged a land exchange to add the 143.3 ha Teton Valley Ranch to the National Elk Refuge (Associated Press, 1986b). In another instance, the National Park Service condemned the McReynolds-Thompson property northwest of Blacktail Butte in Grand Teton National Park to ensure that a subdivision would not be developed on the property (Thuermer, 1986a). Other properties on the edge of the Park have been proposed for development, and several of these have been developed. The nonprofit Jackson Hole Land Trust was organized by local residents to prevent development of private land by securing conservation easements. Individuals granting such conservation easements or making gifts of land may often secure tax advantages. To date the Land Trust has protected 1523 ha of land in Jackson Hole from development although the lands generally can be used for agriculture and ranching.

Two popular ski resorts have been developed in Jackson Hole. With development of these ski areas came inevitable condominiums, hotels, shops, and sports developments which occupy limited winter and spring range for elk. The skiing industry benefits the economy of Jackson Hole by dampening the marked seasonal fluctuation in business (Merrifield & Gerking, 1982). Hopefully, any future developments can avoid areas potentially used by wildlife.

Timber industry

The major timber industry affecting Jackson Hole is based in Dubois and Afton, Wyoming. Current proposals for management of the Bridger–Teton National Forest reportedly threaten the financial security of the Louisiana Pacific, Inc. sawmill in Dubois (Prevedel, Robison & Iverson, 1985). The company maintains that it must have 16 million board feet of timber per year to remain solvent. Louisiana Pacific makes up about 25% of the economy of Dubois (Prevedel, Robison & Iverson, 1985). Details on logging in Jackson Hole are presented in Chapter 6.

Oil and gas

At the present time, 31 oil and gas leases are being considered in the Mount Leidy Highlands for exploratory drilling. Although these permits are only for exploratory drilling, the objective of this drilling is obvious, and potential development of oil and gas fields in the Mount Leidy Highlands could lead to substantial conflicts with elk. This area is a key migration corridor for elk en route to the National Elk Refuge and the Gros Ventre winter range, and is also on the spring migration route and includes several areas known to be calving grounds. Forest Service policy on oil and gas leasing is discussed in Chapter 6.

Nonconsumptive values

Principal nonconsumptive activities involving elk in Jackson Hole are simply watching and photographing them. These opportunities exist for the greatest number of people at the National Elk Refuge during winter, and Grand Teton National Park during summer. In recent years the elk reduction program in the Park, which focused on antlerless elk, has provided exceptional opportunities for summertime viewing and photographing of mature bulls in valley areas of the Park, although this has not been an objective of the reduction program.

Ticket sales at the sleigh ride concession and sampling of the number of vehicles and people per vehicle that stop at visitor information facilities indicate that over 5 300 000 people visited the National Elk Refuge during 1973–85 (Figure 9.6). Most of these visitors to the Refuge stop briefly to view wildlife and to read visitor information signs, but visitation to ride sleighs amongst the winter feedground herd at the sleigh ride visitor center is substantial, as summarized in Table 9.2.

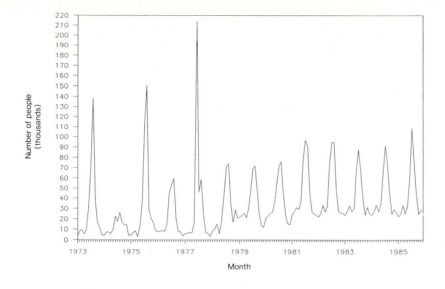

Figure 9.6 Visitation to the National Elk Refuge by month, 1973–85.

Elk in the Greater Yellowstone Ecosystem

The scenic splendor of the Teton Range and the proximity of Jackson Hole to Yellowstone National Park (Figure 9.7) offer a background for high environmental concern and a desire to preserve the Jackson Hole region in a natural state. Two environmental groups are particularly interested in protecting the scenic, recreational, and wildlife values in the Jackson Hole region: the Jackson Hole Alliance for Responsible Planning and the Greater Yellowstone Coalition.

The Greater Yellowstone Coalition's Ungulate Policy Statement (unpublished) expresses concern about the fact that many elk and other ungulates in Yellowstone and Grand Teton National Parks must winter outside Park boundaries. Consequently, developments outside the Parks can substantially impact ungulate populations that summer within the Parks.

Four of the Coalition's policy statements are particularly relevant to the Jackson elk herd:

(1) The Coalition strives to prevent further loss of winter range, migration routes, and birthing areas in all areas surrounding Yellowstone and Grand Teton National Parks.

(2) It is recommended that livestock use on winter ranges be reduced, and specifically they propose that the Wyoming Game and Fish Department explore with Bridger–Teton National Forest and local

Table 9.2. *Sleigh ride concession seasonal summary, National Elk Refuge.*

Season	Adults	Children (6–12)	Children (0–6) free	Schools/groups free/discount Special	Visitor total	Ticket prices adult/child	Expenses	Income
1965–66	1895	558	est. 500		2593	$1.00/0.50	$5159	$2291
1966–67	3125	734	1607		5466	$1.00/0.50	$4729	$4638
1967–68	2618	665	821		4104	$1.50/0.75	$5437	$4426
1968–69	3905	807	1111		5823	$1.50/0.75	$3816	$6510
1969–70	3670	712	1110		5492	$1.50/0.75	$5586	$6039
1970–71	3825	602	748	193	4828	$1.50/0.75	$2272	$5572
1971–72	3190	536	616	261	4603	$1.50/0.75	$2915	$5449
1972–73	6277	1248	1585	164	9274	$1.50/0.75	$5125	$10515
1973–74	4554	701	887	88	6230	$1.50/0.75	$5355	$7444
1974–75	5794	813	726	34	7367	$1.50/0.75	$6518	$9326
1975–76	6934	833	206	530	8553	$2.00/1.00	$9418	$15616
1976–77	4130	488	488	150	5256	$2.00/1.00	$7198	$8921
1977–78	15790	2323	2330	275	20718	$2.00/1.00	$17020	$34111
1978–79	14324	1855	1855	322	18356	$2.00/1.00	$16204	$30931
1979–80	12104	1368	1368	429	15269	$2.00/1.00	$14081	$25688
1980–81	11520	1483	1483	133	14619	$2.50/1.50	$23389	$31325
1981–82	14783	1845	1845	333	18806	$2.50/1.50	$34659	$40960
1982–83	18405	2347	2472	699	23923	$3.50/2.50	$52852	$71997
1983–84	17798	2048	2092	786	22724	$3.50/2.50	$50797	$69559
1984–85	17857	2118	2118	517	22610	$3.50/2.50	$55689	$69253

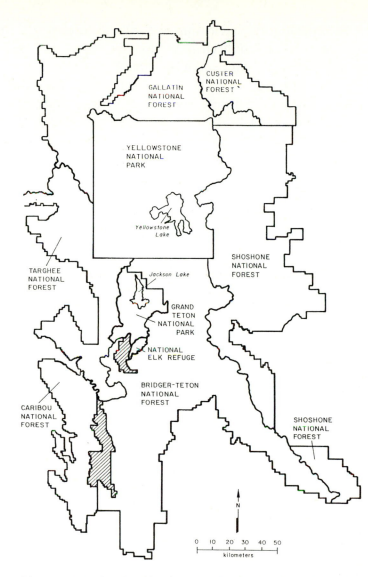

Figure 9.7 Federal and private lands (hatched) in the Jackson Hole region and the Greater Yellowstone Ecosystem. The region contains the largest area of pristine public lands in the contiguous United States.

landowners ways to winter elk in the Gros Ventre River drainage without supplemental feeding.

(3) 'Oil and gas leasing and exploration should be prohibited where subsequent field development will adversely affect elk migration and calving areas.' Specifically, they call for protection of the Mount Leidy Highlands in Bridger–Teton National Forest, recommending

prohibition of oil and gas leasing and large-scale logging in this area which contains important elk migration routes and calving grounds.

(4) Bison numbers in Grand Teton National Park should be controlled because of problems which they create on private lands and on the National Elk Refuge during winter feeding operations. No explicit recommendations on herd size were suggested, but rather this limit is thought best determined by a joint recommendation from the National Park Service, the US Fish and Wildlife Service, and the Wyoming Game and Fish Department.

An ecosystem perspective on elk management

Enlightened resource management must take broader focus than simply managing for one species, for example, elk. Rather, all components of the ecosystem must be considered. Sometimes this seems to be a formidable objective, given the difficulty of understanding the dynamics of even one species, let alone the entire ecosystem. An approach sometimes employed by the US Forest Service is to identify key indicator species, whose viability reflects more broadly on the condition of the ecosystem in general (Thomas, 1979). For Bridger–Teton National Forest, elk have been suggested as a useful indicator species. The implication is that, if management can successfully maintain viable elk populations, key habitats and other wildlife are also probably faring well.

The degree of monitoring and research required to correctly manage an entire ecosystem is staggering. We hope that understanding of the area's ecology will gradually improve so that more sophisticated inputs can guide management decisions. Clearly an integrated systems approach can offer unique insights (Pellew, 1983; Roelle & Auble, 1983).

Interactions between elk and other species

Community ecologists seem to delight in describing intricate higher-order interactions among species of organisms (e.g., Price, Slobodchikoff & Gaud, 1984). One might argue that, ultimately, every species in a community is linked to every other one in a web of complexity that boggles the mind. In reality, however, only a few species usually make much difference (Schaffer, 1982). Elk can be considered a keystone species in the Jackson Hole ecosystem because they play a major role in the ecology of several communities, such as high elevation herblands, aspen forests, riparian willow habitats, and virtually any habitats near a feedground.

Concern has been expressed about consequences of elk browsing to avian communities (Beetle, 1979). Heavy browsing by elk tends to reduce the understory and nesting cover for songbirds (Flack, 1976; DeByle, 1985b). Although this may result in losses of preferred species, for example, bluebirds, avian species diversity may actually be higher in heavily browsed deciduous forests (Casey & Hein, 1983). Consequences of elk grazing or browsing to species diversity of other groups of animals and plants has not been studied to my knowledge.

A recent book by Alston Chase (1986) highlights the potential negative consequences that high elk numbers might have upon the abundance of several herbivores, for example, beaver (*Castor canadensis*), mule deer, moose, and white-tailed deer. Although Chase's book is purely speculation without quantitative evidence for support, his discussions explicitly introduce the potential negative consequences of allowing elk populations to become too large.

Competition with other ungulates

In Chapters 4, 6 and 8 we have discussed the importance of competition between cattle and elk in the Gros Ventre and areas of Grand Teton National Park. On summer ranges, forage is usually abundant, and competition with cattle is not serious, although elk are displaced from areas grazed by cattle. But as noted by several authors (Smith, 1961; Lyon & Ward, 1982; Nelson, 1982), competition from cattle for forage on winter range can seriously reduce the carrying capacity of the area for elk.

Elk and mule deer are known to compete under some circumstances, particularly when cattle share the same range (Mackie, 1985). However, I am not aware of any instance where competition between these species is a management concern within the boundaries of the Jackson elk herd. Mule deer are common on the National Elk Refuge in winter, and are frequently observed on Miller Butte and along slopes on the east boundary of the Refuge.

Bighorn sheep summer range occurs in the Gros Ventre Mountains, and in the Teton Range. Because few elk occupy higher elevation habitats in the Teton Range (W. Barmore, pers. comm.), there is no apparent interaction between elk and the Teton Range sheep population. Bighorn sheep from the Gros Ventre Range winter in the Gros Ventre River valley and immediately adjacent to the National Elk Refuge. Competition between elk and bighorn sheep, particularly during winter, has been suggested by several researchers including Picton (1984), Buechner (1960), Beetle (1962), and Oldemeyer, Barmore & Gilbert (1971). Casebeer (1961) states that bighorn sheep have suffered sharp declines in areas of Jackson Hole where they were in marked competition with elk for winter range, although he presents no data to substantiate this claim. Counts of bighorn sheep adjacent to the National Elk Refuge (Table 9.3) show no apparent trends in sheep numbers since 1961.

Whereas elk selectively browse aspen, moose may enhance aspen stands because they may selectively browse subalpine fir (*Abies lasiocarpa*), which may encroach aspen stands (DeByle, 1985a: 116). Yet, moose also suppress aspen, and the combined effects of elk and moose on aspen stands may be severe (DeByle, 1985a).

Competition between elk and moose is most likely on winter range, particularly in riparian habitats (Stevens, 1974); however, habitat preferences on winter range usually differ (Houston, 1968; Singer, 1979; Nelson, 1982). Moose are primarily browsers and prefer riparian willow habitats. Elk are primarily grazers and prefer grasslands and grass/shrub habitats in northwest Wyoming. Overlap in resource use and spatial distribution is greatest during severe winters when elk use more browse (Jenkins & Wright, 1988).

McMillan (1953) concluded that elk in Yellowstone National Park were forced onto moose range due to overpopulation, and that overbrowsing of willows by elk would eventually reduce moose populations in parts of Yellowstone National Park. Likewise, summer use of riparian willow stands by elk in Grand Teton National Park may reduce winter forage for moose (Martinka, 1969). Combined forage use by cattle and elk in the Spread Creek drainage was

Table 9.3. *Bighorn sheep east of the National Elk Refuge, mostly in Curtis Canyon (G. Roby, unpublished).*

Date of count	Ewes	Lambs	Lambs/100 ewes
12 March 1961	17	8	47
March 1964	51	18	35
29 January 1965	44	20	45
31 October 1966	37	23	62
February 1968	10	5	50
February 1969	39	15	38
9 December 1975	18	11	61
January 1976	24	7	29
6 January 1977	18	5	28
Jan-Feb. 1978	22	5	23
Feb-March 1979	18	7	39
March 1980*	23	8	35
March 1981	20	10	50
January 1982	18	13	72
January 1983	13	7	54
January 1984	16	8	50
January 1985	24	4	17
January 1986	26	15	58

*In 1981, 11 sheep were transplanted from Whiskey Mountain near Dubois to Curtis Canyon. Included were 4 rams, 5 ewes and 2 lambs.

found to be in competition with moose that winter there (Houston, 1968: 43–4). As implied by Houston (1968), however, because most elk concentrate on feedgrounds during winter in Jackson Hole, there is almost certainly less competition between the two species than would exist if elk were not fed.

Predation on elk

Insufficient data exist on elk/wolf interactions to be certain that wolf predation ever limited elk numbers in the region. Most evidence suggests that wolves do not deplete their prey populations (Mech, 1970; Carbyn, 1974). In Jasper National Park in Canada, even in the face of wolf predation, elk numbers appear to be food limited (Carbyn, 1974; Stelfox, 1976). Likewise, although elk are preferred prey of wolves in Riding Mountain National Park in Manitoba, wolf predation was inadequate to offset recruitment (Carbyn, 1983). Ecological efficiency of the wolf trophic level on Isle Royale in Lake Superior is only 1.3%, whereas it is 11.9% for moose (Colinvaux & Barnett, 1979), thereby showing that wolves capture but a small portion of available energy.

Attitudes of wildlife biologists towards predators have changed markedly through this century. In early years, they thought predator control was consistent with sound wildlife management. Then, during the middle part of the century, the importance of predators in some situations was recognized, to the degree that predator control became taboo (Errington, 1967). Recent studies of wolf predation on caribou and moose in Alaska and Canada (Wolfe, 1975; Rausch & Hinman, 1977; Bergerud, Wyatt & Snider, 1983) and on deer in Minnesota and British Columbia (Mech & Karns, 1977; US Fish & Wildlife Service, 1984) have led to the realization that wolves may limit the abundance of prey, particularly when predation by humans supplements that due to wolves. Unfortunately, dynamics of elk/wolf interactions are inadequately understood to predict the role that wolves might have played in limiting elk numbers.

Wolves are not the only potential predators on the Jackson elk herd, although they were probably the most important (Weaver, 1979b). Other predators include cougars (*Felis concolor*; Murie, 1951c; Hornocker, 1970; Cunningham, 1971) and grizzly bears (*Ursus arctos*; Cole, 1972; Martinka, 1978), although numbers of both of these species are so low that neither could have a significant influence on elk numbers. Black bears (*Ursus americanus*) prey substantially on elk calves in some areas (Schlegel, 1976); although such observations have not been reported for Jackson Hole. Black bears also occasionally kill adult elk (Barmore & Stradley, 1971). Coyotes (*Canis latrans*) are known to kill elk calves (Houston, 1982), but usually function as a scavenger on elk carcasses (Houston, 1978; Weaver, 1979a).

Wolves in Jackson Hole?

The proposal to reintroduce wolves to Yellowstone National Park (US Fish & Wildlife Service, 1984) has met strong objection from local livestock interests and consequently from the Wyoming Congressional Delegation. Yet, over 90% of visitors to Yellowstone National Park favor reintroduction of wolves into the Park (Skinner, 1985). To foster better acceptance of the proposal to reintroduce wolves, during the summer of 1985 Yellowstone National Park hosted an exhibit entitled 'Wolves and Man' at the Grant Village Visitor Center. Likewise, the Wyoming Game and Fish Department apparently supports wolf reintroduction into Yellowstone National Park (Skinner, 1985), although their official policy opposes the reintroduction until an experimental reintroduction is attempted elsewhere.

If wolves are successfully reintroduced into the Yellowstone region, they will probably eventually disperse southward into Grand Teton National Park, the Teton Wilderness, and perhaps to the National Elk Refuge. One can only conjecture the effect that wolves will have on large winter concentrations of elk on the Refuge and other feedgrounds. They might disperse elk into surrounding forest lands. Yet, herding is thought to be an anti-predator defense strategy (Geist, 1982). The Recovery Plan states that if predation on big game herds is in 'significant conflict with State wildlife management agency objectives, wolf control that would not jeopardize wolf recovery would be considered'.

The area proposed for wolf reintroduction includes Teton Wilderness, Grand Teton National Park, and 'adjacent public lands' (US Fish & Wildlife Service, 1984). Under the Recovery Team's plan, wolves dispersing into areas designated for livestock grazing may be captured or killed. Capture collars will enable researchers to recapture wolves at will.

At the present time, the Wyoming Congressional delegation is unanimously opposed to the wolf reintroduction plan, largely because of strong objections from the livestock industry. Partly because of this political opposition, the Directors of the National Park Service and the US Fish and Wildlife Service are not proceeding with the preparation of an Environmental Impact Statement which is necessary before wolves could be released. Until wolves are reintroduced, however, the Greater Yellowstone Ecosystem will be incomplete. Because of the winter feeding programs and extensive culling on the Jackson elk herd, the void left by extermination of the wolf will be of less consequence than for other less-intensively managed herds in Yellowstone National Park.

Summary
1. Elk contribute substantially to the economy of Jackson Hole. Outfitted hunting alone is estimated to generate over $4 million annually. Guides and outfitters in the region are very active in promoting elk management for consumptive use.
2. Although difficult to quantify, elk offer important nonconsumptive use value to Jackson Hole. Elk viewing opportunities are an integral part of the tourist industry.
3. Greatest threats to the future of the Jackson elk herd occur from potential oil and gas mining and timbering in the Mount Leidy Highlands where such developments could alter migration routes to winter ranges (see Chapter 6).
4. Because of their great numbers, elk play a major role in the ecology of the Greater Yellowstone Ecosystem. Because winter ranges have been preempted by man, natural population processes cannot be completely restored for the Jackson elk herd. Developments on private land within the Jackson herd unit continue to usurp elk winter range.
5. Wolves were exterminated from the Jackson–Yellowstone region during the 1920s by government predator control programs. Proposals to reintroduce wolves to the Yellowstone region have met strong opposition from livestock interests and the Wyoming Congressional delegation despite general public sentiment favoring the reintroduction.

Chapter 10

Conclusions and evaluation

Thus far, we have reviewed various issues of key management significance for the Jackson elk herd and have analyzed data which bear upon these issues. This concluding chapter is a sounding board to summarize the key findings and to highlight management implications. In addition to consequences for management, such a research effort always uncovers more questions than it answers, and I endeavor to rank research needs for the future, as well as to evaluate which monitoring data should be continued.

Despite the great complexity of managing the Jackson elk herd, I think that a few points are clear:

Historical (Chapter 1)
(1) Despite an extensive winter feeding program for more than 75 years, the Jackson elk herd is generally healthy and viable and provides extensive opportunities for recreational hunting and nonconsumptive uses.
(2) The Jackson Hole Cooperative Elk Studies Group is a model of interagency cooperation for research and management of a wildlife population.

Migration and seasonal distribution (Chapter 2)
(3) Interchange between the Jackson elk herd and surrounding herds is substantially less than 10% and appears to play a minor role in the population dynamics of the Jackson elk herd.
(4) Timing and spatial distribution of elk migration is altered substantially from year to year by variation in weather, primarily snow depth. This strongly affects the ability of hunters to harvest elk from some segments of the herd.

(5) Differential hunting pressure has substantially reduced herd seg-ments which migrate through eastern portions of the Jackson herd unit, specifically in thè Togwotee area. Increased kill in these areas can be attributed to improved access for hunters afforded by logging roads constructed during the 1960s. Whereas prior to 1960, the Togwotee migration was the largest for the herd, hosting 20–30 % of the migration, it is now one of the smallest migration routes with less than 5% of the migratory elk.

(6) Attempts to restore historical distribution of migration (Cole, 1969), have been unsuccessful because spatial distribution of hunter harvest has been inconsistent with attainment of this goal.

The elk population (Chapter 3)

(7) Attendance of elk at winter feedgrounds varies with winter weather; therefore, counts at feedgrounds yield biased indices of population size. At least 50% of the variation in feedground counts can be attributed to winter weather.

(8) Most of the variation in population size of the Jackson elk herd is attributable to recruitment of calves and hunter kill.

(9) Calf recruitment and survival of calves to yearlings is density-dependent. Limited evidence also suggests that dispersal of yearlings is also density-dependent.

Habitat ecology (Chapter 4)

(10) There is no evidence that the Jackson elk herd has caused deteriora-tion of its habitat except on or near winter feedgrounds.

(11) Although elk have contributed to the decline of many aspen stands in Jackson Hole, particularly in the upper Gros Ventre River drainage, the decline is partly attributable to fire suppression during the past several decades.

Winter feeding (Chapter 5)

(12) The National Elk Refuge is a showplace of wildlife management performing a critical role by replacing limited elk winter range previously usurped by land development and offering spectacular opportunities for viewing wildlife.

(13) Total overwinter losses of elk on the National Elk Refuge average only 1.2%, which is less than the 1.8% annual loss rate for adult cattle on ranches in Wyoming.

(14) Feeding at the National Elk Refuge and the Gros Ventre

feedgrounds appears to be efficiently managed and adequate to sustain calf recruitment.

(15) Brucellosis occurs in approximately 30% of elk in the Jackson elk herd. Approximately one half of all infected cow elk abort their first calf or bear dead young. Brucellosis can be transmitted between elk and cattle.

Managing the Forest (Chapter 6)

(16) Perhaps the greatest opportunities/threats for future management of the Jackson elk herd exist on Bridger–Teton National Forest. Key management tools are (a) constraint over oil and gas exploration and mining to minimize potential impacts of roads and development, (b) regulation of timber management and harvest to benefit or to minimize impacts on elk distribution and abundance, (c) management for road closures to control human disturbance and hunter access, (d) control of cattle grazing, and (e) implementation of prescribed burning programs for habitat enhancement.

(17) There is inadequate control of cattle distribution on Gros Ventre elk winter range. As a consequence, some areas are severely overgrazed during most years.

Wyoming Game and Fish Department (Chapter 7)

(18) The Wyoming Game and Fish Department established a quota of no more than 7500 elk wintering on the National Elk Refuge and 2182 at three state feedgrounds in the upper Gros Ventre River valley. These quotas are maintained by regulations controlling hunter harvest.

(19) During the past 20 years, the Jackson elk herd appears to have been managed at levels near to maximum sustained yield.

Grand Teton National Park (Chapter 8)

(20) The elk reduction program in Grand Teton National Park has helped to secure balanced harvests from the Jackson elk herd to ensure a check on population size. Kills within the Park and on the Refuge during the past 10 years have removed primarily elk that summer in the Park.

Public interests (Chapter 9)

(21) Wildlife values are extremely important in the economy of Jackson Hole, and wildlife is an integral part of the attraction for tourists to the area.

(22) Continuing community expansion and subdivision development in Jackson Hole still threatens remaining winter range for the Jackson elk herd.

Data needs

Effective management of the Jackson elk herd must be based upon adequate information about the status of the herd and its habitat. More data exist for the Jackson elk herd than for most wildlife populations in the western United States because of a mandate included in section 6 of Law 787 requiring that such data be collected. I believe that most of these data should continue to be collected to ensure that any future changes are adequately documented. It is extremely important that these data be obtained with consistent methodology so that data are comparable from year to year. Methodology should be thoroughly documented to ensure consistent data collection, even though personnel change.

Data currently obtained by agencies represented on the Jackson Hole Cooperative Elk Studies Group are listed below and are ranked in order of their significance to future management of the Jackson elk herd. My ranking is partly subjective, although based in part on the quality and reliability of the data.

(1) Harvest estimates based upon mail surveys. Exact procedures employed have varied from year to year, but procedures and changes in them have not been documented. This needs to be done in order to standardize harvest estimates consequent to refinements in sampling or analysis.
(2) Collection of counts and sex and age structure information of elk on feedgrounds should be continued. I do not encourage extensive efforts to estimate the number of elk wintering away from feedgrounds because a reliable methodology for such estimates has not been developed.
(3) Total and per capita quantities of supplemental feed provided to elk should be estimated to monitor the effectiveness of feeding programs. Instantaneous count techniques can improve estimates of days of elk use on the Refuge and Gros Ventre feedgrounds (Appendix J).
(4) Counts and data on sex and age structure should be collected annually from valley portions of Grand Teton National Park. Every effort should be made to obtain counts each summer because a missing year results in two years for which annual changes, for example, dN/dt, cannot be monitored.

(5) Track counts should be conducted at least twice weekly from mid-October through mid-December. Track counts are necessary to document the effectiveness of management efforts attempting to restore historical migration patterns.

(6) Ear tags should be placed on elk handled for various purposes, although I cannot justify continued emphasis on tagging large numbers of elk, at least until a thorough evaluation of tag returns has been accomplished.

In addition to these data, I think that better monitoring of vegetation will ensure that managers continue to give attention to habitat ecology for the Jackson summer ranges. In particular, the Red Creek vegetation transects in southern Yellowstone National Park show trends in utilization which should be verified by further monitoring. Other established transects and browse use plots in Grand Teton National Park should be rechecked at least every 2–3 years. In addition to the newly instituted range trend transects, established vegetation sampling and range use transects (Yorgason, 1963) should be monitored in the upper Gros Ventre winter range areas to ensure proper distribution of cattle use in the area. Unfortunately, proper evaluation of range requires large sample sizes because variance is often large. Adequate evaluation will require much greater commitment than has been given in the past. A compromise solution might be to support a new MS-level graduate student every 5–10 years to update and evaluate range surveys on elk summer and winter ranges.

It is recognized that many of the data collection tasks outlined here can be very time consuming. With increasing responsibilities for agency personnel during recent years, biologists sometimes find that there is inadequate time to perform a proper job of data collection. Agency supervisors must recognize the extensive field work required to obtain sound data for basing management decisions, and either enlist additional support or reduce other time demands on field biologists.

Research needs

There are enormous gaps in our knowledge of the population dynamics and ecology of the Jackson elk herd, some of which hinder efficient management. Many of these gaps can be filled by research, and some are of immediate significance to management. Again, I will rank these according to my subjective prioritization.

(1) Research is urgently needed on dispersal of calves and yearlings. The most practical methodology is probably radio-telemetry using

expandable collars such as employed by Kuck, Hompland, & Merrill (1985). In my opinion, this research is fundamental to gaining an understanding of population processes regulating the Jackson elk herd and it demands highest priority as a research need.

(2) Further study of methods for describing winter severity is required to better predict winter feedground attendance. I recommend that a year-round weather station be established in the upper Gros Ventre River drainage, for example, at the Goosewing Ranger Station where some weather data have been collected. Variables which should be given attention include temperature, precipitation, wind velocity, snow packing and crusting. Surveys to locate elk and their winter ranges, particularly during mild winters, should validate the model at equation 3.2. However, methodologies for censusing elk off feedgrounds are unreliable and do not merit extensive effort.

(3) A study similar to the one performed by Martinka (1965, 1969) should be conducted to evaluate and refine methodology for estimating the number of elk summering in forested areas of Grand Teton National Park. Use of the change-in-ratio technique developed by Paloheimo & Fraser (1981) to estimate population size should also be explored. Detailed information on the age composition of the harvest should be recorded for 5 years at the Dubois check station, and for elk killed in the Park and on the Refuge.

(4) Research is needed to develop a model for determining the optimal time to initiate feeding. Existence of such a model need not determine management action, but it could be a useful management tool. Data inputs will require measures of winter severity, and indices of condition for the animals. Some relevant data on body mass dynamics may be available from elk collected during feeding trial studies.

(5) To complement previous studies (Thorne & Butler, 1976; Smith & Robbins, 1984; Oldemeyer, Robbins & Smith, unpublished), research is needed to construct an energetics model for optimizing feeding rations. Again, inputs include measures of winter severity, but also availability and quality of natural forage. Carefully controlled experimental work will be required similar to the feeding trial studies at the Sybille Research Station near Laramie (Thorne, Dean & Hepworth, 1976). The objective is to extend the feeding trial studies to assess the importance of environmental variables on dynamics of body mass and fat in elk. Much background information has already been compiled but not followed through to its logical completion.

(6) Research on elk–livestock interactions should be synthesized so that the Forest Service could incorporate cattle grazing into their habitat effectiveness models (Boss *et al.*, 1983; Leege, 1984). In addition, characterization of habitats along migration routes should be studied so that these habitats could eventually be incorporated into habitat effectiveness models.

(7) Methods for monitoring the distribution and sex and age composition of elk on summer ranges, particularly in southern Yellowstone National Park and the northern Teton Wilderness, should be researched. Seasonal employees may be used to collect data once a viable sampling scheme is devised. Continuation of the pellet group transects and vegetation samples initiated by Gruell (1973) seems desirable, although to be meaningful, larger sample sizes will be required (Neff, 1968).

Management recommendations

Research needs listed above are inseparable from and complement the following management recommendations, which derive from my analysis and interpretation of data compiled on the Jackson elk herd. These recommendations and priorities reflect a high value given to wildlife. Agency officials that are responsible for making management decisions must balance wildlife with other resources, consider legal and policy constraints that differ amongst the agencies, and they must consider numerous economic, social and political factors that can strongly influence their management programs. Hopefully there will be some middle ground that is optimal for elk, elk habitats and as many people as possible.

(1) The Park hunt should be continued, although experimental reductions in quotas may be attempted to refine estimates of the yield curve presented in Figure 3.33. Effectiveness of the elk herd reduction program could be improved by opening certain lands west of the Snake River to hunting, and by increasing the flexibility of the quota system to allow for increased quotas during mild years when harvests are not obtainable from Forest lands. The recent focus on harvesting antlerless elk should be discontinued because of the resulting extraordinary bull/cow ratios. I view continuation of the Park hunt as possibly necessary but undesirable. However, it is unnecessary to continue harvest practices that create unnatural demography in the elk herd.

As an ecologist with a fundamental interest in population

dynamics, I would very much like to learn the consequences of experimentally reducing or terminating the elk reduction program in Grand Teton National Park, as proposed by the National Park Service (Grand Teton National Park, 1985). What might happen if the Park hunt was reduced or terminated? The yield curve at Figure 3.33 suggests that the combined consequence of density-dependent recruitment, mortality and dispersal is adequate to stabilize elk numbers at approximately 2000 elk in the central valley population. However, we cannot be certain of the population response at high numbers until an experimental manipulation is conducted. The northern Yellowstone herd has not been culled within Yellowstone National Park since 1969 (Houston, 1982), yet it is not clear that the population has yet stabilized.

To ensure an acceptable number of huntable elk outside the Park, any increases in the Park population will require increasing the allowable quota of elk to be fed on the National Elk Refuge. If some of the density dependence among Park elk is yearling dispersal into surrounding areas, an even greater number of elk may need to be harvested from the Jackson elk herd to keep the population in check. Without the Park hunt to ensure harvests (see Chapter 8), during some years hunter kill may be inadequate to secure necessary kill and increased culls on the National Elk Refuge may be required to keep the herd within quotas.

I do not recommend closure of the Park hunt. To my mind, the most compelling arguments for continuation of Park hunts are (a) closure would be irreversible given strong public sentiments against hunting in National Parks (Wood, 1984), and (b) closure would hamper efforts to sustain adequate harvest for the purpose of limiting numbers of elk on winter feedgrounds and for restoring historical distributions.

One solution might be to simply terminate both the Park hunt and supplemental feeding at the Refuge. I am confident that the public will find such a solution unacceptable. Given that most natural winter range is now inaccessible to the elk herd due to human interference, during many winters the remaining natural winter range could support considerably fewer elk than we now enjoy.

(2) Restrictions on hunter kill should be retained on eastern segments of the Jackson elk herd to restore historical distribution of migration. Until eastern elk migrations have been restored, any necessary increases in harvest should come from Grand Teton National Park and the National Elk Refuge.

(3) There must be improved enforcement and administration of cattle distribution on the Gros Ventre elk winter range. Lack of supervision by permittees of cowboys working allotments frequently results in poor distribution of cattle. Hiring more cowboys or trucking cattle to summer pastures may help to ensure proper rest-rotation grazing on the Fish Creek Allotment. Current allotments appear to be designed to minimize competitive interaction between cattle and elk on the Gros Ventre winter range, but failure to enforce grazing permits has led to serious conflicts. The Jackson Hole Cooperative Elk Studies Group should evaluate existing transects to ensure adequate monitoring of trends in forage use and range condition.

(4) All existing grazing exclosures in Jackson Hole should be maintained annually to ensure that they continue to exclude elk, deer and livestock. The ecology of aspen and plant succession in general remains poorly understood. Research on these topics requires exclosures as baseline controls. Cole (1971) argued that data from exclosures could be misinterpreted and recommended that they be dismantled in Yellowstone National Park, a view which was contested by Barmore (1968). The risk of misinterpreting data from exclosures can be countered with informed ecological interpretation. Without such data, correct interpretations are less likely. Granted, ecological interactions are complex and may be difficult to understand. But surely, resource management based upon ignorance cannot be sound.

(5) Cattle grazing in the Uhl Hill area of Grand Teton National Park and in the Gros Ventre River valley should be postponed until 30 June of each year to reduce the risk of transmitting brucellosis between elk and cattle. Value of this recommendation is reinforced by known sensitivity of bluebunch wheatgrass to grazing during the flowering period in June (Mueggler, 1975). Ranchers should be encouraged to vaccinate all calves against brucellosis.

(6) The Dubois check station should sample hunters throughout the entire hunting season. Continuous operation of the check station during this time is probably more expensive than the data warrant. But rather than shortening the period of operation at either or both ends of the season, the station should be closed at randomly selected and unannounced times during the hunting season.

(7) When possible, road closure on the National Forest should be permanent and year round for all parties including administrative

access. Increased road closures should be attempted, particularly in the Spread Creek drainage. Public reaction to road closures is often negative once people become accustomed to gaining access to an area. However, if new roads are never opened for public access, this reaction can be circumvented.

(8) Wildlife habitat should be considered at early stages of planning timber sales on Bridger–Teton National Forest. The Forest's elk habitat effectiveness model should be employed early in the planning process to ensure optimal areas for logging, rather than attempting to correct the impacts of proposed timber sales.

(9) Herd unit objectives for recruitment and bull ratios should be lowered to realistic levels, for example, 30 calves/100 cows and 15 bulls/100 cows. I do not support the recent reduction in harvest objectives for the Jackson elk herd. These changes were clearly a reaction to public pressure to rebuild herd numbers.

(10) Current efforts should continue to enhance forage production on the National Elk Refuge, and to secure private lands for expansion of the Refuge and Grand Teton National Park when critical winter ranges are threatened with development. As Madsen (1985) emphasized, 'However well intentioned, winter feeding programs will never make up for the loss of winter habitat.'

(11) I recommend extensive prescribed burning of elk ranges on National Forest lands, certainly at much higher levels than have been practiced by the Forest in the past. Small burns are ineffective and do not achieve desired results. Unfortunately, specific guidelines for optimal size of burns have not been formulated. Mixed species stands are the ones which should receive highest priority for burning, to reduce conifer encroachment on stands of aspen or poplar. Herbicides or cutting may be viable alternatives to burning for stands containing inadequate fuels to support fire (DeByle & Winokur, 1985). It is a reasonable long-term objective to reduce the extent and frequency of winter feeding on Gros Ventre feedgrounds.

(12) Oil and gas development and extensive timbering should be banned from the Mt Leidy Highlands. This is a critical elk migration route where disturbance could easily cause a shift in migration from Gros Ventre feedgrounds to the National Elk Refuge. Perhaps most sensitive is the Sohare Creek and Gunsight Pass area where Amoco Oil Company has plans for an oil well.

Because of enormous incentives for development in Jackson Hole, I suspect that land management will remain the greatest challenge for managers of the Jackson elk herd. It is tragic that a national treasure enjoyed each year by millions of people may be degraded for the pecuniary rewards of the oil and gas industry. Continuing pressures for economic development will certainly present many difficult problems in the future, yet, it would be wrong to end this book on anything but an optimistic note. The Jackson elk herd is one of the finest wildlife resources in the world. I am confident that it will retain this status for many years to come.

APPENDIX A.
ANALYSIS OF TRACK COUNTS

In this appendix I examine the role of hunter kill and snow depth in determining patterns in year-to-year variation in the weighted relative proportion of track counts (C'_i). To begin, first-order differencing was performed to remove trends (Bloomfield, 1976). Conceptually, this is simply the change between weighted relative proportions of tracks from one year to the next, i.e.,

$$\Delta C'_i = C'_i - C'_{i-1} \tag{A.1}$$

Next, the variation in $\Delta C'_i$ is examined.

Records of hunter kill of elk from Grand Teton National Park and the National Elk Refuge are maintained by the National Park Service and the US Fish and Wildlife Service. Hunter harvest from each of 12 special management areas (Figure 3.23) is sampled annually at the Dubois check station (see Chapter 3). For each year, I computed the total kill from each special management area as follows:

$$A_j = [(\text{JACKSON}) - (\text{PARK} + \text{NER})] \, [a_j / \textstyle\sum a_j] \tag{A.2}$$

where JACKSON = total kill from the Jackson herd unit estimated by mail surveys, PARK and NER = recorded kill of elk from Grand Teton National Park and the National Elk Refuge, a_j is the number of elk checked at the Dubois check station from the jth special management area, where $j = 1, 2, \ldots, 12$.

I calculated numerous multiple regression models to characterize the influence of harvest and snow depth on migration patterns. Snow depth data from the Moran weather station was again used since this weather station is nearest to the track transects. All variables were screened prior to analysis for skewness and kurtosis and transformations were performed where appropriate to normalize distributions. The general model is of the form

$$\Delta C'_i = \beta_0 + \beta_j A_j + \beta_k S_k + \epsilon \tag{A.3}$$

for transect i and various A_j's, with $j = 1, 2, \ldots, 12$, and S_k's are monthly maximum depth of snow for $k =$ October, November, December.

Although many of these models were intriguing, with eight transects and a minimum of 15 independent variables, the total number of statistically significant models is rather staggering. Therefore, I used multidimensional scaling (Kruskal, 1964a,b) to reduce the number of dependent variables. Results suggest that transects within the following two sets of transects are structurally similar and may be pooled: (1) Snake River, Pacific Creek and Buffalo River transects (PACTRACK), and (2) Boundary, Blackrock and Four-Mile Meadow transects (BLKTRACK; Figure A.1). This grouping is more easily accepted since transects within these sets are adjacent to one another geographically. Thus the eight transects may be reduced to these two sets plus the Burned Ridge and Togwotee transects (which are not similar) for analysis.

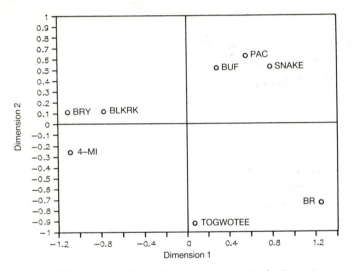

Figure A.1 Multidimensional scaling of track counts from each of the eight transects. The two clusters of transect routes reflect similar patterns of variation in the weighted proportion of tracks crossing these transects.

In Table A.1 I present multiple regression models for these four transect groups. Many of the patterns documented by these statistical models support the conclusion that hunter harvest has had a role in shaping the spatial distribution of elk migration. For example, some elk that migrate across the Togwotee transect subsequently migrate across the Spread Creek harvest areas (Anderson, 1958), and as expected, harvests from these areas are inversely correlated with Togwotee transect counts in subsequent years.

Several of the independent variables seem rather unassociated with track counts on the various transects. This is expected because the weighted proportion of counts is influenced by variation occurring over all of the transects. For example, the positive correlation of harvests on special management area 8 (Spread Creek) with subsequent track counts on the Burned Ridge transect simply reflects that removal of animals from Spread Creek will increase the relative proportion of elk in the Burned Ridge transect during the following year, even though the animals actually using the Burned Ridge transect are unaffected.

I also employed regression on a principal component to simplify the complex interactions between variables in this data set. I calculated the first major axis through the changes in weighted relative proportions of track counts for the eight transects. The correlations between the eight transects and this first 'track factor' (Figure A.2) reflect increased numbers of elk crossing Park transects, but fewer elk crossing more easterly transects (except Togwotee). This corresponds well with the patterns described by multidimensional scaling, since PACTRACK is positively correlated with the 'track factor' ($r = 0.636$), whereas BLKTRACK is highly negatively correlated ($r = -0.98$). Next I used this single 'track factor' as a dependent variable to calculate a multiple regression model with five harvest and climate variables (Figure A.2).

December maximum snow depth (S_{Dec}) is inversely correlated with the 'track factor'. Since the 'track factor' correlated positively with migration across Park transects but negatively with that for Forest transects, this indicates that heavy snows result in a greater proportion of elk crossing Boundary,

Table A.1. *Multiple regression models characterizing the role of hunter harvest and snow depth in the change in weighted proportions of elk tracks crossing highway transect routes, 1950–84. For partial correlations,* *$P < 0.05$, **$P < 0.01$.

△ Burned Ridge $= 0.068$ (Park**$_{t-2}$) $- 0.085$ (Park**$_{t-1}$) $- 1.29$ (A_2**) $+ 15.9$ (log A_8**) $+ 63.8$ (log A_{12}**) $+ 31.7$ (log K_1**) $+ 0.013$ (K_7) $- 0.078$ (Gros*) $- 638.6$
$r = 0.913$; 8, 12 d.f.; $F = 7.474$; $P = 0.001$

△ Togwotee $= 0.014$ (Park**) $- 0.006$ (Spread*) $- 0.183$ (S_{Nov}**) $+ 5.75$ (log K_1**) $+ 4.117$ (log S_{Oct}**) $- 40.98$
$r = 0.913$; 5, 16 d.f.; $F = 15.95$; $P < < 0.001$

△ PACTRACK $= 5.914$ (log A_4) $- 6.5$ (log S_{Oct}*) $- 26.4$
$r = 0.528$; 2, 20 d.f.; $F = 3.87$; $P = 0.038$

△ BLKTRACK $= 0.364$ (S_{Dec}*) $+ 0.228$ (A_6**) $- 10.8$ (log A_1) $- 0.156$ (A_{11}*) $- 16$
$r = 0.69$; 4, 18 d.f.; $F = 4.09$; $P = 0.016$

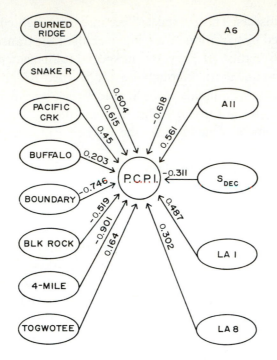

Figure A.2 Path diagram of a principal component regression. The center represents a track factor with correlations with original track count variables at the left. Loadings suggest that the track factor reflects increasing tracks crossing Park transects while inversely correlated with transect counts east of the Park. The independent variables on the left are linked to the track factor with partial correlations. A6 and A11 are hunter kill from special management areas 6 and 11 (see Figure 3.23) during the previous season. LA1 and LA8 are logarithms of kill from areas 1 and 8. S_{Dec} is the maximum depth of snow on the ground at Moran, Wyoming in December.

Table A.2. *Pearson product-moment correlation coefficients between the change in the weighted relative frequency of tracks crossing each of eight transects versus maximum depth of snow on the ground at Moran, S_i.*

	$\log (S_{Oct} + 1)$	S_{Nov}	S_{Dec}
Burned Ridge	0.097	0.055	−0.119
Snake River	−0.218	−0.117	−0.083
Pacific Creek	−0.411*	−0.246	−0.327
Buffalo River	−0.321	−0.061	−0.104
Boundary	−0.081	0.002	0.282
Blackrock	−0.000	0.404*	0.349
Four-Mile Meadow	0.283	0.275	0.225
Togwotee	0.333	−0.407	−0.209

*$P < 0.05$

Blackrock and four-Mile Meadow track transects and a lower proportion crossing Park transects. This interpretation is reinforced by the pattern in the bivariate correlation coefficients (Table A.2). Although only two of these correlations are statistically significant, consistent direction of sign reinforces the pattern.

APPENDIX B.
EVALUATION OF VALLEY COUNTS OF ELK IN GRAND TETON NATIONAL PARK

During the winter of 1962–63, 183 elk on the National Elk Refuge were marked with colored neckbands, and during the winter of 1963–64, another 779 elk were color-banded. By intensive observation of elk during the summer of 1964, a minimum of 96 of these neck-banded elk were identified summering in the valley areas of Grand Teton National Park (Martinka, 1969). Counting routes were established where elk were systematically counted and classified by sex and age class during early morning and evening in late July and early August. Based upon reobservations of the 96 neck-banded elk, a Lincoln-Petersen estimate of population size was calculated from maximum counts over these established observation routes.

Six observation areas were originally established by Martinka (1965, 1969) and counts of elk have been conducted on these areas every year since 1963 except 1983: (1) Willow Flats, a point-count location near Jackson Lake Lodge, (2) Signal Mountain, another point-count location from a vantage point at the top of the south end of Signal Mountain, (3) Inside Road, between Moose and Jackson Lake Dam and then east to Moran, (4) Potholes, along a dirt road from Signal Mountain to the River Road northeast of Burned Ridge, (5) Baseline Flats, south of Burned Ridge along the River Road, and (6) Timbered Island, now observed from a dirt road accessed along the River Road, but when Martinka (1969) counted it, from a road that encircled Timbered Island. In 1964, Martinka (1969) also counted 75 elk at Uhl Hill, but this area was not counted any other year; therefore the 1964 count from Uhl Hill was eliminated from all subsequent calculations.

Each area is counted at least twice, and key areas such as the Potholes may be counted as many as four times. Although Martinka originally conducted counts both morning and evening, evening counts have been eliminated since 1973, because maximum counts were never obtained then. All counts are completed by 08:00 hr. The maximum for each area is used as an index of relative abundance (R. P. Wood, unpublished).

Because mature bulls were less likely to be observed, they were not included in mark-recapture population estimates, but were later added to the population estimate based upon the proportion of adult males in a classified sample of elk (e.g., 11% adult males in 615 elk classified in 1964; Martinka, 1969). For 1964, the population of females, calves, and yearling males was estimated using the model

$$\hat{N} = n \cdot M / m = (557)(96)/47 = 1138 \tag{B.1}$$

where $\hat{N}$ is the estimated number of females, calves and yearling males, M is the known number of elk with bands, n is the summed maximum counts over all counting routes, and m is the number of marked animals observed during these maximum counts. This estimator is identical to dividing the number of marked elk present by the proportion of animals observed that bore neck-bands.

Currently, maximum trend counts are divided by 0.45 to derive estimates of the central valley population (Cole, 1969). The exact source for the 0.45 observability correction is not clear, but an approximate value can be reconstructed from data in Martinka (1969). He observed 557 elk (excluding mature bulls) in the central valley during trend counts in 1964, and used a Lincoln-Petersen method to estimate a population of 1129 elk (again excluding mature bulls). This yields an observability correction of 0.493. However, his estimated population size does not include 43 elk counted at Moran Meadows (Willow Flats), which he simply added to the estimated value for the central valley. Furthermore, none of these values include mature bulls, which Martinka notes are more difficult to observe during summer and therefore, the observability correction is adjusted downward to 0.45 (Cole, 1969).

I have found discrepancies in reported values for the estimated central valley population for 1963 and 1964. Cole (1969) reported the trend counts to be 643 and 632 for 1963 and 1964 respectively, whereas Martinka (1969) indicates that these counts exclude mature males. Martinka (1969) separately estimates the number of mature males based upon composition counts and adds these to his estimates for cows, calves and spike bulls. To obtain numbers comparable to later years, I have recalculated estimates of central valley populations for 1963 and 1964 using Martinka's (1969) Table 2 data:

> 0.45(central valley counts + Willow Flats counts) + estimated
> number of mature bulls (B.2)

To reconstruct trend counts including mature bulls for these years I multiplied the estimated populations by 0.45. These estimates (Table 3.2) differ from

Martinka's (1969) but the methodology is at least consistent with Cole (1969) and with the methodology which I used for all later years.

Because the bull segment seems to be poorly accounted for, I have also calculated estimates for the non-bull segments of the population using Martinka's original observability correction of 0.493. Estimated population of females, calves and yearling males is compared with estimates for the total central valley population in Figure 3.6 for 1969–85. Differences in trends depicted by the two estimates reflect the increasing proportion of mature bulls in valley counts during recent years.

Criteria for evaluating a population estimator include precision and accuracy (White *et al.*, 1982). Precision of the estimator may be evaluated by calculating confidence intervals to surround estimated population sizes. I use both Seber's (1970) and Bailey's (1951, 1952) methods to calculate variance in the estimates of $\hat{N}$. Seber's variance estimate is derived from Chapman's (1951) approximately unbiased estimator based upon a hypergeometric distribution:

$$\hat{N} = \frac{(M+1)(n+1)}{(m+1)-1} \tag{B.3}$$

with variance

$$\text{Var}(\hat{N}) = \frac{(M+1)(n+1)(M-m)(n-m)}{(m+1)^2(m+2)} \tag{B.4}$$

Bailey's model is derived from a binomial approximation to the hypergeometric distribution

$$\hat{N} = \frac{M(n+1)}{(m+1)} \tag{B.5}$$

with variance

$$\text{Var}(\hat{N}) = \frac{M^2(n+1)(n-m)}{(m+1)^2(m+2)} \tag{B.6}$$

Because Cole's (1969) 0.45 adjustment employs different methodology than the usual Lincoln-Petersen estimator, a confidence interval cannot be precisely estimated. But the confidence interval for the calf-cow-yearling male segment can be estimated and then expanded by 20% to account for lower observability of males. Although somewhat arbitrary, this method allows estimation of the approximate magnitude of confidence intervals around population estimates. Incorporating the 20% adjustment for bulls, I calculate the confidence intervals as follows

$$\text{CI}_{(95\%)} = \hat{N} \pm 2.16[\text{Var}(N)]^{\frac{1}{2}} \tag{B.7}$$

For 1964, the Chapman–Seber model yields 1127 ± 258, and the more conservative Bailey model gives 1116 ± 359. These extrapolated confidence intervals for each year (Table 3.2) necessarily assume that observability and sampling intensity are constant in all years. Needless to say, these assumptions are probably violated.

Although these variance estimators are widely used, I emphasize that they both assume proportions, i.e., n is fixed and not a random variable. In reality, Lincoln-Petersen estimates are usually ratio estimates with random variables in both numerator and denominator. However, true estimates of variance for a ratio would require replicated sampling each year and are therefore impractical for most applications. Nevertheless, Seber (1973) argues that for large $\hat{N}$, the variance estimate will be very similar for models assuming fixed n or random n.

These confidence intervals are about 20–30% of the estimates, which seems reasonably precise. However, precision does not imply accuracy, and valley estimates are probably somewhat biased. At least three considerations may bias these estimates, and in each instance cause them to underestimate actual population size for the central valley: (1) not all marked individuals in valley areas may have been observed (Martinka, 1969), (2) during summer, mature bulls are more difficult to observe than other elk (Martinka, 1969) and Cole's 4.3% adjustment in the population correction factor may not have adequately accounted for unobserved bulls. This is most serious in recent years where the proportion of mature bulls in the valley population has become very high indeed. And, (3) present coverage at Timbered Island is less extensive because the road surrounding the base of Timbered Island has been closed (R. P. Wood, pers. comm.).

Unfortunately, without a study of the scope of Martinka's, I am not aware of any way to rigorously evaluate these possible sources of bias. Bruce Smith has recently analyzed data from the National Elk Refuge's radio-telemetry study and notes that 48% of 97 radio-collared elk summered in Grand Teton National Park. This is substantially more than in Martinka's study where only 12.3% (96/779) elk adorned with neck-bands summered in central portions of the valley. For either of these percentages to be representative, however, it is essential that elk were sampled from throughout the Refuge herd so that each individual had an equal probability of being sampled.

Some of the year-to-year variation in central valley counts may be attributable to summer precipitation, since during dry summers, succulent forage may be most available within forested habitats where observability is low (Robert P. Wood, pers. comm.). This hypothesis is consistent with observations by Irwin & Peek (1983a, b) and Marcum & Scott (1985) who have shown that local patterns of summer distribution are largely correlated with forage density and that weather patterns have a major effect on habitat selection. However, I am unable to reject the null hypothesis that central valley trend counts were not correlated with precipitation in June, July or the summed precipitation for these two months (P > 0.1). Correlation coefficients are all less than 0.212 and two are negative.

Clearly more research is needed to reliably estimate the number of elk using Grand Teton National Park versus other areas. Central valley counts are probably biased indicators of population trends, because Martinka (1969) emphasized that observability for adult males is lower than for other sex and age classes. Therefore, because the proportion of adult males in valley counts has increased substantially over years that trend counts have been conducted ($r_s = 0.779$, $n = 15$, $P < 0.001$), population estimates for more recent years may be more biased (underestimates) than in earlier years. The extent of this bias is unknown, however. Yet this bias has no bearing on estimates of the female, yearling male and calf segment of the population.

Another independent verification for valley estimates is to show correlation with other population parameters. Curiously, there is no correlation between the valley counts and the number of elk counted at the National Elk Refuge the following or previous winter ($r = 0.05$ for previous winter, $r = 0.176$ for following winter, $P > 0.1$ for both cases). Likewise, there is no correlation between valley estimates and Refuge counts standardized for a tough winter ($r = 0.163$, $n = 19$, $P > 0.1$). However, the cow-calf-yearling male segment of the central valley counts is positively correlated with counts of elk on the Refuge during the subsequent winter ($r = 0.53$, $n = 15$, $P < 0.05$). Also, there is a positive correlation between valley estimates and elk mortalities on the National Elk Refuge during the following winter ($r = 0.498$, $n = 20$, $P < 0.05$), which corroborates nicely with Smith's (1985) observation that scabies-induced mortality in bulls is highest when bulls are most abundant.

Barmore (1984) noted a weak inverse correlation ($0.05 < P < 0.1$) between the harvest of elk from Grand Teton National Park and valley counts the following summer. Updating his analysis to include 1985 data, the pattern is now statistically significant ($r = -0.435$, $n = 22$, $P = 0.043$) when 1965 and 1966 are included in the analysis, but not significant if 1965 and 1966 are deleted from the analysis ($r = -0.28$, $n = 20$, $P = 0.233$). Cole (1969) suggests that disturbance of elk by a control program for forest insects during 1965 and 1966 resulted in poor counts in those years, thereby justifying deleting them from the analysis. Also, central valley population estimates and kill rate the previous autumn are inversely correlated ($r = -0.497$, $P < 0.025$), where kill rate = the number of elk killed in the Park and on the Refuge divided by the preceding summer's valley count. I find kill rate discomforting, however, because counts at $t - 1$ are in the denominator of kill rate, therefore possibly ensuring an inverse correlation. This should only be a serious difficulty if valley counts are autocorrelated, however, and the first lag autocorrelation for valley counts is only 0.234 ($P > 0.1$). Nevertheless, it is reassuring that the actual total kill of elk from the Park is correlated with subsequent valley counts. Confirmation of these expected patterns supports valley counts as a trend indicator.

In a final attempt to assess the validity of the valley counts as indicators of population trend, I calculated linear multiple regression models of the form:

$$N = \beta_0 + \beta_1 X_1 + \beta_2 X_2 + \ldots + \beta_i X_i + \epsilon$$

where X_i's are population variables hypothesized to contribute to population fluctuations in elk. Specifically, for independent variables I used (1) hunter kill in the Park and on the Refuge during the previous season, (2) hunter kill in the Park and Refuge during the previous two seasons (Picton, 1984), (3) hunter kill from the entire Jackson herd unit, (4) counts of elk on the National Elk Refuge during the following and (5) preceding winters, (6) recruitment of calves estimated from composition counts on the Refuge during the previous winter, (7) mortality on the Refuge, and (8) summer precipitation values. I even resorted to stepwise procedures in a desperate attempt to find some combination of these variables that could predict the valley estimates. I attempted to fit these models with all years included and again after deleting 1965 and 1966 justified by the disturbances to the elk during those summers (Cole, 1969). Except for the patterns described above, none of the models that I postulated were statistically significant.

APPENDIX C. CENSUS AND CLASSIFICATION OF ELK ON THE NATIONAL ELK REFUGE, 1912–86

Year	Date	Mature bulls	Spike bulls	Cows	Calves	UNCL	On feed	Off feed	Total
1912									7250
1913									4000
1914									6150
1915	no feeding								
1916									8000
1917									6000
1918									10000
1919									3000
1920									8000
1921									3500
1922									4300
1923									3400
1924									4800
1925									5500
1926	no feeding								
1927		450	258	3650	1143		5521	1104	6625
1928									7500
1929									6000
1930									7000
1931		131	41	2425	513		3110		3110
1932		1160	194	4305	1214		6873	1527	8400
1933									7460
1934	no feeding								
1935		644	315	6142	1860		8961	539	9500
1936		605	25	3020	432		4083	117	4200
1937									4000
1938		768	264	4552	1071		6655		6655
1939							7637		

Appendix C. (*cont.*)

Year	Date	Mature bulls	Spike bulls	Cows	Calves	UNCL	On feed	Off feed	Total
1940	no feeding								9500
1941	19 Feb	1182	530	5876	2216	520	10324	676	11000
1942	13 Mar	1177	530	5947	2187	9841	9841	1344	11185
1943	15 Jan					9779	9779	1921	11700
1944	no feeding								
1945	26 Feb	661	245	3522	798		5226	1217	6443
1946	07 Mar	833	311	3929	1168		6241	1759	8000
1947	not classified								
1948	not classified								
1949	25 Feb	1070	588	3639	1572	1190	8059	1364	9423
1950	not classified					8700	8700	1000	9700
1951	not classified					9500	9500		
1952	07 Feb	1221	480	4502	1148		7351	1196	8547
1953	14 Feb					9000	9000	200	9200
1954	25 Feb	1155	577	5000	1383		8115	1415	9530
1955	15 Jan								8000
1956	12 Mar	1489	447	7362	1719	595	11612		11612
1957	Jan					6800	6800	1200	8000
1958						5695	5695	1305	7000
1959	09 Mar	1020	236	3204	1224	363	5684	496	6543
1960	29 Feb	996	302	2413	1035		4746	1030	5776
1961	13 Jan	696	343	3683	869		5591	1114	6705
1962	03 Jan	1070	531	4191	1874		7666	556	8222
1963	03 Jan	614	365	4086	762		5827		
1964	26 Feb	967	307	4867	1775		7916		
1965	28 Jan	930	594	4889	1533		7946		
1966		803	383	4127	1243		6556		
1967	10 Feb	822	407	4622	1518		7369	296	7665
1968	15 Feb	689	429	4465	1076		6659	150	6809
1969	31 Feb	886	573	5523	2223		9205		
1970	19 Feb	864	533	5428	1596		8421	775	9196
1971	02 Mar	736	580	5181	1557		8054	823	8877
1972	04 Feb	720	532	4781	1582		7615	835	8450
1973	28 Feb	740	416	4745	1293		7194	286	7480
1974	28 Feb	716	557	5233	1372		7878	678	8556
1975	03 Mar	745	522	4768	1415		7450		
1976	05 Mar	980	511	4725	1643		7858	515	8373
1977	15 Mar	616	459	3511	1146			5732	
1978	23 Feb	1393	424	5073	1523	82	8143	396	8891
1979	23 Feb	1503	544	4347	1434	130	7828	594	8552
1980	04 Mar	1680	441	4443	1185	25	7749	206	7980
1981	no feeding							6300	

Appendix C. (*cont.*)

Year	Date	Mature bulls	Spike bulls	Cows	Calves	UNCL	On feed	Off feed	Total
1982	24 Feb	1261	405	3801	1063	216	6530	489	7235
1983	24 Feb	1118	488	3312	960	45	5878	346	6269
1984	23 Feb	1073	345	2886	706	45	5010	311	5366
1985	22 Feb	984	293	3500	981	106	5758	402	6266
1986	20 Feb	819	350	4039	1222		6430	296	6726

Notes:

'On feed' is the total number of elk on feed.

'Off feed' are elk wintering immediately adjacent to the NER.

'Total' is the sum of these plus any off feed on NER (UNCL).

APPENDIX D. POST-SEASON ELK CLASSIFICATION AT GROSS VENTRE FEEDGROUNDS, INCLUDING ALKALI, FISH CREEK AND PATROL CABIN

Year	Mature bulls No.	%	Spike bulls No.	%	Cows No.	%	Calves No.	%	Total
1960	18	2.9%	19	3.1%	476	77%	105	17%	618
1961	3	0.4%	7	0.9%	594	77%	162	21%	771
1962	10	1.4%	19	2.6%	473	66%	216	30%	718
1963	6	0.7%	8	1.0%	682	85%	107	13%	803
1964	25	2.5%	37	3.6%	707	69%	249	24%	1018
1965	10	0.9%	44	4.1%	727	68%	281	26%	1062
1966	47	2.7%	41	2.3%	1216	70%	441	25%	1745
1967	53	3.2%	55	3.4%	1205	74%	319	20%	1632
1968	17	0.8%	84	3.9%	1533	72%	493	23%	2127
1969	66	2.0%	97	3.0%	2353	73%	708	22%	3224
1970	47	2.0%	71	3.0%	1703	73%	513	22%	2334
1971	64	2.3%	156	5.5%	1962	70%	634	23%	2816
1972	40	1.7%	186	7.9%	1587	67%	549	23%	2362
1973	45	2.8%	68	4.2%	1191	74%	313	19%	1617
1974	24	1.4%	91	5.3%	1234	71%	380	22%	1729
1975	29	2.1%	106	7.8%	971	71%	260	19%	1366
1976	48	3.4%	62	4.4%	1037	74%	263	19%	1410
1977	no feeding this year – no classifications done								
1978	70	4.1%	43	2.5%	1151	68%	439	26%	1703
1979	31	1.8%	85	5.0%	1174	70%	396	23%	1686
1980	69	3.4%	78	3.9%	1358	68%	504	25%	2009
1981	no feeding this year – no classifications done								
1982	43	2.1%	44	2.1%	1467	70%	541	26%	2095
1983	8	0.6%	40	2.8%	1049	73%	345	24%	1442
1984	25	1.6%	52	3.3%	1152	73%	345	22%	1574
1985	15	1.1%	28	2.1%	1044	79%	241	18%	1328
24-year mean	31	1.9%	63	3.9%	1169	72%	367	22%	1633

APPENDIX E. HUNTER KILL OF ELK FROM THE JACKSON HERD UNIT

Year	Total kill	Park and Refuge kill	Bulls	Spikes	Cows	Calves	Unknown
1950	3638	0	1256		1283	479	620
1951	4810	184	1469		1970	597	774
1952	2108	237	771		823	257	257
1953	2823	179	957		878	288	700
1954	2594	104	1113		1014	287	180
1955	3207	310	1234		1383	465	125
1956	3106	325	1028		1576	502	0
1957	2252	160	713		884	280	375
1958	3057	110	920	330	1318	489	0
1959	1542	0	502	201	631	201	7
1960	2074	0	767		689	288	330
1961	2639	278					2639
1962	1320	270	531	153	436	118	82
1963	2996	720	1203	405	914	370	104
1964	2975	803	1025	509	1021	352	68
1965	3250	787	1115	660	1004	393	78
1966	3226	745	1275	632	1022	297	0
1967	2974	506	1342	537	879	216	0
1968	3090	588	1082	694	1067	221	26
1969	3444	716	1102	771	1226	345	0
1970	5680	990	1693	925	2491	564	7
1971	4245	654	1263	345	1985	598	54
1972	4086	578	1424	242	1907	513	0
1973	4684	648	1886	22	2075	701	0
1974	2774	500	1060	33	1326	355	0
1975	3526	840	1121	510	1443	452	0
1976	1429	293	455	215	576	183	0

Appendix E (*cont.*)

Year	Total kill	Park and Refuge kill	Bulls	Spikes	Cows	Calves	Unknown
1977	3756	862	1188	611	1705	252	0
1978	2880	886	744	387	1425	324	0
1979	3321	971	810	415	1566	530	0
1980	3740	1005	1256	591	1481	412	0
1981	4290	974	1288	676	1853	473	0
1982	3548	1193	846	497	1600	605	0
1983	2355	935	731	434	884	306	0
1984	1524	504	788	299	344	93	0
1985	1368	536	923	86	293	66	0
Means	3065	539	1054	430	1228	368	179

Year	Area 1	Area 2	Area 3	Area 4	Area 5	Area 6	Area 7	Area 8	Area 9	Area 10	Area 11	Area 12	Park and Refuge	Total checked
1950	20	41	796	544	29	415	0	100	299	47	50	222	93	3035
1951	28	49	909	1268	14	150	1	481	185	82	79	225	70	3963
1952	38	23	154	124	38	218	0	41	193	34	85	216	136	1758
1963	33	42	288	330	27	203	0	67	268	17	93	232	403	2003
1964	8	24	170	43	14	99	84	176	82	12	18	36	397	1381
1965	6	25	261	74	30	123	52	424	229	19	27	82	527	1907
1966	5	20	198	120	17	78	72	25	283	35	47	142	474	1516
1967	14	8	103	276	28	72	53	52	357	40	42	111	293	1448
1968	4	5	113	197	12	59	83	40	375	28	62	180	343	1501
1969	10	13	100	236	11	76	160	33	480	26	76	225	411	1857
1970	9	3	103	260	20	121	174	85	678	266	135	284	632	2770
1971	7	18	136	237	10	90	115	46	434	87	93	194	424	1891
1972	4	14	73	149	40	46	9	69	343	57	72	174	297	1343
1973	2	5	76	771	37	74	8	168	605	60	136	258	389	2589
1974	0	12	62	30	36	63	12	59	225	21	74	179	187	960
1975	8	4	66	290	22	94	22	198	295	82	79	222	479	1861
1976	7	7	17	38	26	39	1	26	169	3	29	106	115	583
1977	3	7	114	143	32	108	6	70	338	34	63	143	377	1438
1978	1	7	33	77	25	45	12	230	6	44	115	134	167	896
1979	7	13	100	135	17	98	14	28	350	28	88	129	310	1317
1980	4	8	53	91	24	107	11	20	287	17	100	153	149	1024
1981	2	2	22	58	8	69	4	10	207	16	69	206	93	766
1982	5	6	93	137	14	50	31	77	246	37	70	168	595	1529
1983	3	14	86	52	6	43	2	17	123	4	26	190	435	1001
1984	4	11	25	43	9	42	0	7	52	3	41	158	307	702
Means	9	15	166	229	22	103	37	102	284	44	71	175	324	1642

APPENDIX G. ELK MORTALITIES ON THE NATIONAL ELK REFUGE, BEGINNING WINTER 1940–41

Year	Classified Mortalities					Unclassified	Total mortalities	Elk on Refuge	Percent mortality
	Bull	Spike	Cow	Calf	Total				
1941	13	6	41	74	134	166	300	11000	2.7%
1942							400	11185	3.6%
1943	84		94	813	991	66	1057	11700	9.0%
1944	16		13	7	36		36	6000	0.6%
1945								6441	
1946	17		17	35	69	40	109	8000	1.4%
1947	31		18	13	62		62	7000	0.9%
1948	21		10	4	35	6	41	6750	0.6%
1949	33		28	117	178		178	9423	1.9%
1950	21		24	112	157		157	9700	1.6%
1951	28		23	92	143		148	9500	1.5%
1952	60		36	145	241		241	8447	2.9%
1953	27		19	4	50		50	9000	0.6%
1954	22		14	13	49		49	9530	0.5%
1955	15		14	3	32		32	8000	0.4%
1956	20		22	45	87		87	11612	0.7%
1957	33		32	15	80		80	8000	1.0%
1958	24		15	22	61		61	7000	0.9%
1959	36		11	20	67	5	72	6543	1.1%
1960	27		21	10	58		58	5425	1.1%
1961	39		15	18	72		72	6705	1.1%
1962	62		32	88	182		182	8222	2.2%
1963	9		8	5	22		22	5827	0.4%
1964	40		20	19	79		79	7916	1.0%
1965	7	2	24	21	54		54	7946	0.7%
1966	5	0	15	13	33		33	6556	0.5%
1967	7	0	10	24	412		41	7665	0.5%

Appendix G. (*cont.*)

Year	Classified Mortalities					Unclassified	Total mortalities	Elk on Refuge	Percent mortality
	Bull	Spike	Cow	Calf	Total				
1968	19	1	35	19	74		74	6809	1.1%
1969	10	2	13	18	43	11	54	9205	0.6%
1970	17	0	68	27	112	1	113	9196	1.2%
1971	13	3	52	47	115	2	117	8877	1.3%
1972	16	5	30	26	77	2	79	8450	0.9%
1973	5	4	47	17	73	2	75	7484	1.0%
1974	14	3	47	59	123	2	125	8556	1.5%
1975	5	1	26	15	47	3	50	7330	0.7%
1976	26	4	65	36	131	5	136	8373	1.6%
1977	9	0	3	0	12	0	12	5732	0.2%
1978	28	3	60	37	128	5	133	8956	1.5%
1979	41	7	32	75	155	2	157	8552	1.8%
1980	15	2	19	22	58	0	58	7980	0.7%
1981	8	0	1	0	9	2	11	6300	0.2%
1982	33	3	30	26	92	2	94	7235	1.3%
1983	102	6	29	29	166	5	171	6269	2.7%
1984	91	1	54	96	242	3	245	5366	4.6%
1985	43	0	21	2	66	0	66	6266	1.1%
1986	36	4	45	35	120	0	120	6726	1.8%

APPENDIX H. SUMMER VALLEY ELK CLASSIFICATION FROM DIRECT COUNTS, GRAND TETON NATIONAL PARK (R. P. WOOD, UNPUBLISHED)

Year	Total no.	Females	Calves	Spikes	Males	Unclassified
1969	794	204	64	62	42	422
1970	1315	330	105	49	102	729
1971	1037	124	38	30	51	794
1972	765	175	39	33	84	434
1973	899	247	89	22	35	506
1974	1382	129	21	46	83	1103
1975	1323	309	116	57	133	708
1976	947	160	47	29	163	548
1977	1316	166	62	35	274	779
1978	2718	178	83	68	307	2082
1979	1512	362	131	64	377	578
1980	1347	124	41	36	248	898
1981	1744	90	48	18	164	1424
1982	1702	110	54	46	188	1304
1983	No count					
1984	1153					
1985	1331	251	97	38	213	732
1986	791					
1987	1138					

APPENDIX I. TONS OF FORAGE PRODUCED ON THE NATIONAL ELK REFUGE

Management Area	Tons of forage produced											
	1973	1974	1975	1976	1977	1978	1979	1980	1981	1982	1983[a]	1984[a]
Headquarters	720.8	1223.9	1409.8	1012.4	977.2	985.3	733.2	1237.3	1910.1[b]	1544.3	1308.0	1168.8
Nowlin Marsh	6122.3	6664.2	6969.0	6478.2	6253.2	8848.9	2536.0	4237.8[b]	7243.1[b]	5650.9[b]	4144.6	3064.7
Miller Butt	390.6	151.8	171.6	209.7	202.4	186.9	167.8	297.8	235.6	191.3	371.3	1164.9
Ben Goe	312.4	265.3	262.5	346.9	334.9	567.3	322.5	533.5	510.0	459.5	552.1	341.1
Peterson	144.0	228.4	241.3	253.7	244.9	329.2	261.0	460.1	320.6	334.0	324.4	611.7
McBride	197.7	208.4	400.0	279.6	269.9	471.0	187.0	432.9	254.0	235.9	753.4	169.5
Poverty Flats	148.6	113.9	172.2	168.3	162.4	397.8	90.0	274.3	343.0	152.9	443.8	536.0
Chambers	427.5	318.0	626.0	564.4	544.9	1243.0	519.0	354.4	675.9	507.6	346.9	239.8
Pederson	702.3	751.6	842.5	647.3	624.8	548.4	493.0	589.5	760.6[b]	353.5	1107.8	442.7
North End	2649.5	2817.1	2719.4	2964.6	2861.8	2928.5	2970.0	4648.0	5494.4	4141.6	9856.4	14613.8
Totals	11815.7	12742.2	13815.2	12946.0	12476.7	16506.3	8279.5	13065.6	17747.3	13577.5	19208.7	22353.0

Notes:

[a] Starting in 1983, the annual growth from sagebrush, rabbitbrush, willow and other woody shrubs was added to the annual production from grasses and other herbaceous plants. To a large extent, this change in procedure of calculating forage production is responsible for the large increases in the 1983 and 1984 estimates.

[b] Area partially prescribe-burned in the spring prior to green-up.

APPENDIX J
INSTANTANEOUS COUNT METHODS

Elk attendance at feedgrounds is often not continuous, since individuals are free to leave the feedground to forage in surrounding areas. Consequently, counts of elk fluctuate from day to day and may include different animals at various times, especially early and late in the winter season. Daily counts of elk on feedgrounds may be viewed as instantaneous counts, for which unbiased sampling procedures and estimators are known (Robson, 1960; Schreuder, Tyre & James, 1975).

The simplest application of instantaneous count procedures for the Jackson elk herd is to estimate total days of elk use on a feedground. In addition, with independent estimates of the average number of days that individual elk spend on the feedground (say from a radio-telemetry study), the total number of elk using the feedground can be estimated. Finally, instantaneous count models allow estimation of total forage consumption by the herd. Data are not presently available for estimating all of these parameters. To encourage appropriate data collection in the future, I will review parameter and variance estimators for specific application to feedground counts.

First, for a given time period, our objective will be to estimate the total number of days of elk use. Total counts of elk on the feedground are ideally conducted on randomly selected days. In practice, counting may only be possible on some days due to weather conditions. This only presents a biased sample if weather that influences counts also influences attendance of elk on the feedground. Let X_t = counts of elk on the t-th day and $\bar{X} = \sum X_t / n$, where n is the number of days that counts were conducted. The total days of elk use during period T is then simply

$$\hat{D} = T \cdot \bar{X} \qquad\qquad (J.1)$$

with variance

$$\mathrm{Var}(\hat{D}) = T^2 \cdot (s^2/n) \qquad\qquad (J.2)$$

and corresponding standard error

$$\mathrm{SE}(\hat{D}) = T \cdot s/(n)^{\frac{1}{2}} \qquad\qquad (J.3)$$

where s^2 is the variance in the sample of counts.

For example, during November 1984, Bruce Smith counted the following numbers of elk on the National Elk Refuge: $X_2 = 530$, $X_{12} = 2671$, $X_{14} = 3425$, $X_{22} = 4254$, where the subscript is the day of the month. These counts average 2720, and the unbiased estimate of total days of elk use during November is $\hat{D} = (30)\,(2720) = 81\,600$. Variance is $2\,549\,561$ and 95% confidence limits are $34\,656$ and $128\,544$ days of elk use. Confidence intervals are large here because X_i's vary greatly early in the seaon while elk are migrating onto the Refuge.

If the mean number of days spent on the feedground per animal could be estimated, then the number of elk visiting the feedground can be estimated by

$$\hat{M} = \hat{D}/(\text{mean days/animal}) \qquad\qquad (J.4)$$

For illustration, consider that an individual elk spends on average $d = 10$ days on the Refuge, as could be determined by a radio-telemetry study, then

$$\hat{M} = 81\,600/10 = 8160 \qquad\qquad (J.5)$$

elk would have used the National Elk Refuge during November 1984. This will only coincide with maximum counts of elk during November only if all elk happen to be on the Refuge when the maximum count is made.

The estimator for M is a ratio of two random variables, therefore the variance estimator is somewhat tedious to calculate:

$$\mathrm{Var}(\hat{M}) = (D/d)^2 \{[\mathrm{Var}(D)/D^2)] +$$
$$[\mathrm{Var}(d)/d^2] - 2[\mathrm{Cov}(D,d)/Dd]\}$$

where Cov (D,d) is the covariance between D and d. The corresponding standard error is simply

$$\mathrm{SE}(\hat{M}) = [\mathrm{Var}(M)]^{\frac{1}{2}} \qquad\qquad (J.7)$$

If free-ranging elk on the Refuge are consuming, say, 4 kg of forage per day, then the total forage consumed per day is

$$\hat{F} = \hat{M} \cdot C \qquad\qquad (J.8)$$

where C is the forage consumed per day. Here, $\hat{F}$ is the product of two random variables and thereby the appropriate variance estimator is

$$\mathrm{Var}(\hat{F}) = \mathrm{Var}(\hat{M} \cdot C)$$
$$= M^2 \mathrm{Var}(C) + C^2 \mathrm{Var}(M) + \mathrm{Var}(M)\mathrm{Var}(C) \tag{J.9}$$

with standard error

$$\mathrm{SE}(\hat{F}) = [\mathrm{Var}(\hat{M} \cdot C)]^{\frac{1}{2}} \tag{J.10}$$

Again, estimates of C require independent investigation.

REFERENCES

Adams, A. W. (1982). Migration. In *Elk of North America: Ecology and Management*, ed. J. W. Thomas & D. E. Toweill, pp. 301–21. Harrisburg, Pennsylvania: Stackpole Books.

Allred, W. J. (1950). Re-establishment of seasonal elk migration through transplanting. *Transactions of the North American Wildlife Conference*, **15**, 597–611.

Altmann, M. (1952). Social behavior of elk, *Cervus canadensis nelsoni*, in the Jackson Hole area of Wyoming. *Behaviour*, **4**, 116–43.

Amory, C. (1974). *Man Kind? Our Incredible War on Wildlife*. New York: Dell Publishing Company.

Anderson, C. C. (1958). *The Elk of Jackson Hole: A Review of Jackson Elk Studies*. Bulletin 10. Cheyenne: Wyoming Game and Fish Commission.

Anderson, E. W. & Scherzinger, R. J. (1975). Improving quality of winter forage for elk by cattle grazing. *Journal of Range Management*, **28**, 120–5.

Anderson, R. M. & May, R. M. (1985). Vaccination and herd immunity to infectious diseases. *Nature*, **318**, 323–9.

Associated Press. (1986a). Delayed elk feeding program harms Gros Ventre River area ranchers. *Laramie Daily Boomerang* (Laramie, Wyoming), 17 Jan 1986, 6.

Associated Press. (1986b). Land swap proposed to add to Elk Refuge. *Laramie Daily Boomerang* (Laramie, Wyoming), 19 Jan 1986, 4.

Austad, S. N. & Sunquist, M. E. (1986). Sex-ratio manipulation in the common opossum. *Nature*, **324**, 58–60.

Bailey, N. T. J. (1951). On estimating the size of mobile populations from capture–recapture data. *Biometrika*, **38**, 293–306.

Bailey, N. T. J. (1952). Improvements in the interpretation of recapture data. *Journal of Animal Ecology*, **21**, 120–7.

Baker, R. R. (1978). *The Evolutionary Ecology of Animal Migration*. New York: Holmes & Meier.

Barmore, W. J., Jr. (1968). *An Evaluation of the Need for Wildlife Exclosures in Yellowstone National Park*. Moose, Wyoming: United States Department of Interior, National Park Service, Grand Teton National Park.

Barmore, W. J., Jr. & Stradley, D. (1971). Predation by black bear on mature male elk. *Journal of Mammalogy*, **52**, 199–202.

Barmore, W. J., Jr. (1984). *A Synthesis of Information on Elk that Summer in or Migrate through Grand Teton National Park*. Moose, WY: United States Department of Interior, National Park Service, Grand Teton National Park.

Barmore, W. J., Jr. (1986). *Population Characteristics, Distribution and Habitat Relationships of Six Ungulates in Northern Yellowstone National Park*. Mammoth, Wyoming: United States Department of Interior, National Park Service, Yellowstone National Park.

Bartos, D. L. & Mueggler, W. F. (1979). Influence of fire on vegetation production in the aspen ecosystem in Western Wyoming. In *North American Elk: Ecology, Behavior and Management*, ed. M. S. Boyce & L. D. Hayden-Wing, pp. 75–8. Laramie: University of Wyoming.

Bartos, D. L. & Mueggler, W. F. (1981). Early succession in aspen communities following fire in western Wyoming. *Journal of Range Management*, **34**, 315–18.

Beall, R. C. (1974). *Winter Habitat Selection and Use by a Western Montana Elk Herd*. Ph.D Dissertation. Missoula: University of Montana.

Becton, P. (1977). The National Brucellosis Program of the United States. In *Bovine Brucellosis, an International Symposium*, ed. R. P. Crawford & R. J. Hidalgo, pp. 401–11. College Station: Texas A & M University Press.

Beetle, A. A. (1952). *A 1951 Survey of Summer Range in the Teton Wilderness Area*. Unpublished report. Jackson, Wyoming: Bridger–Teton National Forest Files.

Beetle, A. A. (1962). Range survey in Teton County, Wyoming. Part 2. Utilization and condition classes. *Wyoming Agricultural Bulletin*, **400**, 1–38.

Beetle, A. A. (1974a). *Range Survey in Teton County, Wyoming. Part IV–Quaking Aspen*. Agricultural Experiment Station Scientific Monograph, no. 27. Laramie: University of Wyoming.

Beetle, A. A. (1974b). The zootic disclimax concept. *Journal of Range Management*, 27, 30–2.

Beetle, A. A. (1979). Jackson Hole elk herd: A summary after 25 years of study. In *North American Elk: Ecology, Behavior and Management*, ed. M. S. Boyce & L. D. Hayden-Wing, pp. 259–62. Laramie: University of Wyoming.

Bendt, R. H. (1960). *Progress Report of the Jackson Hole and Southern Yellowstone Elk Herd Study*. Moose, Wyoming: Grand Teton National Park.

Bendt, R. H. & Yorgason, I. J. (1961). *Elk Migration and Distribution Studies, Jackson Hole Elk Herd*. Moose and Cheyenne, Wyoming: National Park Service and Wyoming Game and Fish Department.

Bendt, R. H. (1962). The Jackson Hole elk herd in Yellowstone and Grand Teton National Parks. *Transactions of the North American Wildlife Conference*, **27**, 191–201.

Bergerud, A. T., Wyatt, W. & Snider, B. (1983). The role of wolf predation in limiting a moose population. *Journal of Wildlife Management*, **47**, 977–88.

Bergstrom, R. C. (1975). Prevalence of *Dictyocaulus viviparus* infections in Rocky Mountain elk in Teton County, Wyoming. *Journal of Wildlife Diseases*, **11**, 40–4.

Bergstrom, R. C. & Robbins, R. (1979). Lungworms, *Dictyocaulus viviparus*, in various age classes of elk (*Cervus canadensis*) in the Tetons. In *North American Elk: Ecology, Behavior and Management*, ed. M. S. Boyce & L. D. Hayden-Wing, pp. 221–3. Laramie: University of Wyoming.

Berntsen, C. M., Litton, R. B., Jr., Lyon, L. J., Packer, P. E., Rees, P. M. & Schmautz, J. E. (1983). *New Directions in Management on the Bighorn, Shoshone and Bridger–Teton National Forests*. Washington, DC: US Department of Agriculture, Forest Service.

Blevins, A. L. & Bryson, A. M. (1980). Teton County's tourism boom. *Wyoming Issues*, 3(2), 2–5.

Bloomfield, P. (1976). *Fourier Analysis of Time Series: An Introduction*. New York: John Wiley.

Blunt, F. M. (1950). Migration study of the Jackson Hole elk herd. *Wyoming Wildlife*, **14**(2), 25–32.

Bobek, B., Boyce, M. S. & Kosobucka, M. (1984). Factors affecting red deer (*Cervus elaphus*) population density in Southeastern Poland. *Journal of Applied Ecology*, **21**, 881–90.

Boss, A., Dunbar, M., Gacey, J., Hanna, P. & Rath, M. (1983). *Elk–Timber Relationships of West-Central Idaho*. Boise, Idaho: Idaho Department of Fish and Game; US Department of Interior, Bureau of Land Management; and US Department of Agriculture, Forest Service.

Boyce, J. S. (1948). *Forest Pathology*, 2nd edn. New York: McGraw-Hill.

Boyce, M. S. & Sauer, J. R. (1978). Elk distribution and behavior in calving areas. *Annual Report of the University of Wyoming – National Park Service Research Center*, **2**, 15–8.

Boyce, M. S. & Hayden-Wing, L. D. (1979). *North American Elk: Ecology, Behavior and Management*. Laramie: University of Wyoming.

Boyce, M. S. (1984). Restitution of r- and K-selection as a model of density-dependent natural selection. *Annual Reviews of Ecology and Systematics*, **15**, 427–47.

Boyd, R. J. (1978). American elk. In *Big Game of North America: Ecology and Management*, ed. J. L. Schmidt & D. L. Gilbert, pp. 11–29. Harrisburg, Pennsylvania: Stackpole Books.

Bridger–Teton National Forest (1986a). *Bridger–Teton National Forest Land and Resource Management Plan. Draft Environmental Impact Statement*. Jackson, Wyoming: US Department of Agriculture, Forest Service.

Bridger–Teton National Forest (1986b). *Proposed Land and Resource Management Plan of the Bridger–Teton National Forest*. Jackson, Wyoming: US Department of Agriculture, Forest Service.

Brown, C. (1985). *Sand Creek Elk*. Boise, Idaho: Idaho Department of Fish and Game.

Brownie, C., Anderson, D. R., Burnham, K. P. & Robson, D. S. (1985). Statistical inference from band recovery data: A handbook. *US Department of Interior, Fish and Wildlife Service Resource Publication*, no. 156.

Bryant, L. D. & Maser, C. (1982). Classification and distribution. In *Elk of North America: Ecology and Management*, ed. J. W. Thomas & D. E. Toweill, pp. 1–59. Harrisburg, Pennsylvania: Stackpole Books.

Bubenik, A. B. (1982). Physiology. In *Elk of North America: Ecology and Management*, ed. J. W. Thomas & D. E. Toweill, pp. 125–79. Harrisburg, Pennsylvania: Stackpole Books.

Buechner, H. K., Buss, I. O. & Bryan, H. F. (1951). Censusing elk by airplane in the Blue Mountains of Washington. *Journal of Wildlife Management*, **15**, 81–7.

Buechner, H. K. (1960). The bighorn sheep in the United States, its past, present, and future. *Wildlife Monographs*, **4**, 1–174.

Buffalo Ranger District (1973). *Teton Wilderness Plan*. Jackson, Wyoming: US Department of Agriculture, Forest Service, Bridger–Teton National Forest.

Burnham, K. P. & Anderson, D. R. (1979). The composite dynamic method as evidence for age-specific waterfowl mortality. *Journal of Wildlife Management*, **43**, 356–66.

Burnham, K. P., Anderson, D. R. & Laake, J. L. (1980). Estimation of density from line transect sampling of biological populations. *Wildlife Monographs*, **72**, 1–202.

Camenzind, F. J. (1978). *Behavioral Ecology of Coyotes* (Canis latrans) *on the National Elk Refuge, Jackson, Wyoming*. PhD Dissertation. Laramie: University of Wyoming.

Carbyn, L. N. (1974). Wolf population fluctuations in Jasper. *Biological Conservation*, **6**, 94–101.

Carbyn, L. N. (1983). Wolf predation on elk in Riding Mountain National Park, Manitoba. *Journal of Wildlife Management*, **47**, 963–76.

Casebeer, R. L. (1961). Habitat of the Jackson Hole elk as a part of multiple resource planning, management, and use. *Transactions of the North American Wildlife Conference*, **26**, 436–47.

Casey, D. & Hein, D. (1983). Effects of heavy browsing on a bird community in deciduous forest. *Journal of Wildlife Management*, **47**, 829–36.

Caughley, G. (1971). An investigation of hybridization between free-ranging wapiti and red deer in New Zealand. *New Zealand Journal of Science*, **14**, 993–1008.

Caughley, G. (1974). Bias in aerial survey. *Journal of Wildlife Management*, **38**, 921–33.

Caughley, G. (1976). Wildlife management and the dynamics of ungulate populations. In *Applied Biology*, Vol. 1, ed. T. H. Coaker, pp. 183–246. London: Academic Press.

Caughley, G. (1977). *Analysis of Vertebrate Populations*. New York: John Wiley.

Caughley, G. (1979). What is this thing called carrying capacity? In *North American Elk: Ecology, Behavior and Management*, ed. M. S. Boyce & L. D. Hayden-Wing, pp. 2–8. Laramie: University of Wyoming.

Cayot, L. J., Prukop, J. & Smith, D. R. (1979). Zootic climax vegetation and natural regulation of elk in Yellowstone National Park. *Wildlife Society Bulletin*, **7**, 162–9.

Chapin, E. A. (1925). New nematodes from North American mammals. *Journal of Agricultural Research*, **30**, 677–81.

Chapman, D. G. (1951). Some properties of the hypergeometric distribution with applications to zoological censuses. *University of California Publications in Statistics*, **1**, 131–60.

Chase, A. (1986). *Playing God in Yellowstone: The Destruction of America's First National Park*. Boston: Atlantic Monthly Press.

Cheeseman, C. L., Little, T. W. A., Mallinson, P. J., Page, R. J. C., Wilesmith, J. W. & Pritchard, D. G. (1985). Population ecology and prevalence of tuberculosis in badgers in an area of Staffordshire. *Mammal Review*, **15**, 125–36.

Clark, A. B. (1978). Sex ratio and local resource competition in a prosimian primate. *Science*, **201**:163–5.

Clutton-Brock, T. H., Guinness, F. E. & Albon, S. D. (1982). *Red Deer: Behavior and Ecology of Two Sexes*. Chicago: University of Chicago Press.

Clutton-Brock, T. H., Albon, S. D. & Guinness, F. E. (1982). Competition between female relatives in a matrilocal mammal. *Nature*, **300**, 178–80.

Clutton-Brock, T. H., Albon, S. D. & Guinness, F. E. (1984). Maternal dominance, breeding success and birth sex ratios in red deer. *Nature*, **308**, 358–60.

Clutton-Brock, T. H., Albon, S. D. & Guinness, F. E. (1985). Parental investment and sex differences in juvenile mortality in birds and mammals. *Nature*, **313**, 131–3.

Clutton-Brock, T. H., Major, M. & Guinness, F. E. (1985). Population regulation in male and female red deer. *Journal of Animal Ecology*, **54**, 831–46.

Cochran, W. G. (1977). *Sampling Techniques*, 3rd edn. New York: John Wiley.

Cole, G. F. (1963). *Range Survey Guide*, revised ed. Moose, Wyoming: Grand Teton Natural History Association.

Cole, G. F. (1965). *Elk Ecology and Management Investigations*. Moose, Wyoming: National Park Service.

Cole, G. F. & Yorgason, I. J. (1965). *Elk Migration Study, Jackson Hole Elk Herd, 1965*. Moose and Cheyenne, Wyoming: National Park Service and Wyoming Game and Fish Department.

Cole, G. F. (1969). *The Elk of Grand Teton and Southern Yellowstone National Parks*. Research Report. GRTE–N–1, Office of Natural Science Studies, US Department of Interior, National Park Service. (reprinted, 1981). Washington, DC: US Government Printing Office.

Cole, G. F. (1971). Some considerations in the use of enclosures to assess the biotic effects of herbivores and departures from natural conditions in Yellowstone National Park. *National Park Information Paper*, **13**, 1–5.

Cole, G. F. (1972). Grizzly bear–elk relationship in Yellowstone National Park. *Journal of Wildlife Management*, **36**, 556–61.

Cole, G. F. (1983). A naturally regulated elk population. In *Symposium on Natural Regulation of Wildlife Populations*, 10 March 1978, ed. F. L. Bunnell, D. S. Eastman & J. M. Peek, p. 62–81. Moscow, Idaho: University of Idaho.

Colinvaux, P. A. & Barnett, B. D. (1979). Lindeman and the ecological efficiency of wolves. *American Naturalist*, **114**, 707–18.

Collins, W. B. & Urness, P. J. (1982). Mule deer and elk responses to horsefly attacks. *Northwest Science*, **56**, 299–302.

Collins, W. B. & Urness, P. J. (1983). Feeding behavior and habitat selection of mule deer and elk on northern Utah summer range. *Journal of Wildlife Management*, **47**, 646–63.

Cook, R. D. & Jacobson, J. O. (1979). A design for estimating visibility bias in aerial survey. *Biometrics*, **35**, 735–42.

Craighead, F. C., Craighead, J. J., Cote, C. E. & Buechner, H. K. (1972). Satellite and ground radiotracking of elk. In *Animal Orientation and Navigation*, ed. S. R. Galler, K. Schmidt-Koenig, C. J. Jacobs & R. E. Belleville, pp. 99–111. Washington, DC: Scientific and Technical Information Office, National Aeronautics and Space Administration.

Craighead, J. J. (1952). *A Biological and Economic Appraisal of the Jackson Hole Elk Herd*. New York: New York Zoological Society and The Conservation Foundation.

Craighead, J. J., Atwell, G. & O'Gara, B. W. (1972). Elk migration in and near Yellowstone National Park. *Wildlife Monographs*, **29**, 1–48.

Craighead, J. J., Craighead, F. C., Jr., Ruff, R. L. & O'Gara, B. W. (1973). Home ranges and activity patterns of nonmigratory elk of the Madison Drainage herd as determined by biotelemetry. *Wildlife Monographs*, **33**, 1–50.

Croft, A. R. & L. Ellison. (1960). *Watershed and Range Conditions on Big Game Ridge and Vicinity, Teton National Forest, Wyoming*. Ogden, Utah: US Department of Agriculture, Forest Service.

Crowe, D. M. (1981). *Big Game Damage Costs*. Planning Report 5E. Cheyenne: Wyoming Game and Fish Department.

Crowe, D. M. (1983). *Comprehensive Planning for Wildlife Resources*. Cheyenne: Wyoming Game and Fish Department.

Cunningham, E. B. (1971). A cougar kills an elk. *Canadian Field-Naturalist*, **85**, 253–54.

Dalke, P. D., Beeman, R. D., Kindel, F. J., Robel, R. J. & Williams, T. R. (1965a). Seasonal movements of elk in the Selway River drainage, Idaho. *Journal of Wildlife Management*, **29**, 333–8.

Dalke, P. D., Beeman, R. D., Kindel, F. J., Robel, R. J. & Williams, T. R. (1965b). Use of salt by elk in Idaho. *Journal of Wildlife Management*, **29**, 319–32.

Davis, D. S., Booer, W. J., Mims, J. P., Heck, F. C. & Adams, L. G. (1979). *Brucella abortus* in coyotes. I. A seriologic and bacteriologic survey in eastern Texas. *Journal of Wildlife Diseases*, **15**, 367–72.

DeByle, N. V. (1979). Potential effects of stable versus fluctuating elk populations in the aspen ecosystem. In *North American Elk: Ecology, Behavior and Management*, ed. M. S. Boyce & L. D. Hayden-Wing, pp. 13–9. Laramie: University of Wyoming.

DeByle, N. V. (1985a). Animal impacts. In *Aspen: Ecology and Management in the Western United States*, ed. N. V. DeByle & R. P. Winokur, pp. 115–23. US Department of Agriculture, Forest Service. General Technical Report RM–119. Fort Collins: Rocky Mountain Forest and Range Experiment Station.

DeByle, N. V. (1985b). Management for esthetics and recreation, forage, water, and wildlife. In *Aspen: Ecology and Management in the Western United States*, ed. N. V.

DeByle & R. P. Winokur, pp. 223–32. US Department of Agriculture, Forest Service. General Technical Report RM–119. Fort Collins: Rocky Mountain Forest and Range Experiment Station.

DeByle, N. V. (1985c). Wildlife. In *Aspen: Ecology and Management in the Western United States*, ed. N. V. DeByle & R. P. Winokur, pp. 135–52. US Department of Agriculture, Forest Service. General Technical Report RM–119. Fort Collins: Rocky Mountain Forest and Range Experiment Station.

DeByle, N. V. & Winokur, R. P. (1985). *Aspen: Ecology and Management in the Western United States*. US Department of Agriculture, Forest Service. General Technical Report RM–119. Fort Collins: Rocky Mountain Forest and Range Experiment Station.

Dexter, D. (1984). Sixty minutes and the Teton elk. *Wyoming Wildlife*, **48**(5), 3.

Dieterich, R. A. (1981). *Alaskan Wildlife Diseases*. Fairbanks: University of Alaska.

Dirks, R. A. & Martner, B. E. (1982). *The Climate of Yellowstone and Grand Teton National Parks*. US Department of Interior, National Park Service, Occasional Paper No. 6. 26 pp.

Dobson, F. S. (1982). Competition for mates and predominant juvenile male dispersal in mammals. *Animal Behaviour*, **30**, 1183–92.

Eberhardt, L. L. (1985). Assessing the dynamics of wild populations. *Journal of Wildlife Management*, **49**, 997–1012.

Edge, W. D. & Marcum, C. L. (1985). Movements of elk in relation to logging disturbances. *Journal of Wildlife Management*, **49**, 926–30.

Edge, W. D., Marcum, C. L. & Olson, S. L. (1985). Effects of logging activities on home-range fidelity of elk. *Journal of Wildlife Management*, **49**, 741–4.

Edwards, R. Y. & Ritcey, R. W. (1958). Reproduction in a moose population. *Journal of Wildlife Management*, **22**, 261–8.

Errington, P. L. (1967). *Of Predation and Life*. Ames: Iowa State University Press.

Filion, F. L. (1981). Importance of question wording and response burden in hunter surveys. *Journal of Wildlife Management*, **45**, 873–82.

Flack, J. A. D. (1976). Bird populations of aspen forests in western North America. *Ornithological Monograph*, **19**, 1–97.

Fletcher, R. R., Lewis, E. P., Premer, G. E. & Taylor, D. T. (1979). *Recreation and Tourism in the Teton County Economy*. Agricultural Extension Service. Laramie: University of Wyoming.

Flook, D. R. (1970). Causes and implications of an observed sex differential in the survival of wapiti. *Canadian Wildlife Service, Report Series*, **11**, 1–71.

Fowler, C. W. (1981). Density dependence as related to life history strategy. *Ecology*, **62**, 602–10.

Fowler, C. W. (1987). A review of density dependence in populations of large mammals. In *Current Mammalogy*, Vol. 1, ed. H. H. Genoways, pp. 401–41. New York: Plenum.

Fretwell, S. D. & Lucas, H. L., Jr. (1970). On territorial behavior and other factors influencing habitat distribution in birds. I. Theoretical development. *Acta Biotheoretica*, **19**, 16–36.

Frison, G. C. & Bradley, B. A. (1981). Fluting Folsom projectile points: archaeological evidence. *Lithic Technology*, **10**, 13–16.

Gasson, W. (1987). Managing elk the Wyoming way. *Wyoming Wildlife*, **51**(9), 16–25.

Geist, V. (1971). *Mountain Sheep: A Study in Behavior and Evolution*. Chicago: Univ. Chicago Press.

Geist, V. (1982) Adaptive behavioral strategies. In *Elk of North America: Ecology and Management*, ed. J. W. Thomas & D. E. Toweill, pp. 219–77. Harrisburg, Pennsylvania: Stackpole Books.

Gilpin, M. E. & Ayala, F. J. (1973). Global models of growth and competition. *Proceedings National Academy of Sciences, USA*, **70**, 3590–3.

Gladney, G. (1985). Tourism touted as critical to state's future. *Jackson Hole News*, 20 October 1985, 12.

Gordon, I. J. (1988). Facilitation of red deer grazing by cattle and its impact on red deer performance. *Journal of Applied Ecology*, **25**, 1–10.

Grand Teton National Park. (1985). *Natural Resources Management Plan and Environmental Assessment*. Moose, Wyoming: United States Department of Interior, National Park Service.

Greer, K. R. (1966). Fertility rates of the northern Yellowstone elk populations. *Proceedings, Annual Conference of the Western Association of State Game and Fish Commissioners*, **46**, 123–8.

Greer, K. R. & Hawkins, H. W. (1967). Determining pregnancy in elk by rectal palpation. *Journal of Wildlife Management*, **31**, 145–9.

Greer, K. R., Hawkins, H. W. & Cutlin, J. E. (1968). Experimental studies of controlled reproduction in elk (wapiti). *Journal of Wildlife Management*, **32**, 368–76.

Gross, J. E. (1969). Optimum yield in deer and elk populations. *Transactions of the North American Wildlife and Natural Resources Conference*, **34**, 372–87.

Grover, K. E. & Thompson, M. J. (1986). Factors influencing spring feeding site selection by elk in the Elkhorn Mountains, Montana. *Journal of Wildlife Management*, **50**, 466–70.

Gruell, G. E. (1973). *An Ecological Evaluation of Big Game Ridge*. Jackson, WY: US Department of Agriculture, Forest Service, Teton National Forest.

Gruell, G. E. & Loope, L. L. (1974). *Relationships among aspen, fire, and ungulate browsing in Jackson Hole, Wyoming*. Ogden, Utah: US Department of Agriculture, Forest Service, Intermountain Region, and Denver, Colorado: US Department of Interior, National Park Service, Rocky Mountain Region.

Gruell, G. E. & Roby, G. (1976). Elk habitat relationships before logging on Bridger–Teton National Forest, Wyoming. In *Elk-Logging-Roads Symposium Proceedings*, ed. S. R. Hieb, pp. 110–21. Moscow: University of Idaho.

Gruell, G. E. (1979). Wildlife habitat investigations and management implications on the Bridger–Teton National Forest. In *North American Elk: Ecology, Behavior and Management*, ed. M. S. Boyce & L. D. Hayden-Wing, pp. 63–74. Laramie: University of Wyoming.

Gruell, G. E. (1980a). *Fire's Influence on Wildlife Habitat on the Bridger–Teton National Forest, Wyoming, Vol. I – Photographic Record and Analysis*. US Department of Agriculture, Forest Service Research Paper INT–235. Ogden, Utah: Intermountain Forest and Range Experiment Station.

Gruell, G. E. (1980b). *Fire's Influence on Wildlife Habitat on the Bridger–Teton National Forest, Wyoming, Vol. II – Changes and Causes, Management Implications*. US Department of Agriculture, Forest Service Research Paper INT–252. Ogden, Utah: Intermountain Forest and Range Experiment Station.

Guest, J. E. (1971). *Carrying Capacity of Elk Summer Range*. PhD dissertation. Laramie: University of Wyoming.

Habib, J. H. (ed.) (1984). *Wyoming Agricultural Statistics*. Cheyenne: Wyoming Crop and Livestock Reporting Service.

Hansen, C. (1977). *A Report on the Value of Wildlife*. Ogden, Utah: US Department of Agriculture, Forest Service, Intermountain Forest and Range Experiment Station.

Harniss, R. O. & Nelson, D. L. (1984). *A Severe Epidemic of* Marssonina *Leaf Blight on Quaking Aspen in Northern Utah*. US Department of Agriculture, Forest Service Research Note INT–339. Ogden, Utah: Intermountain Forest and Range Experiment Station.

Harper, J. (1969). Relationships of elk to reforestation in the Pacific Northwest. In *Wildlife and Reforestation in the Pacific Northwest*, ed. H. C. Black, pp. 67–71. Corvallis: Oregon State University.

Hart, J. H. (1986). *Relationship among Aspen, Fungi, and Ungulate Browsing in Colorado and Wyoming*. Fort Collins, Colorado: Rocky Mountain Forest and Range Experiment Station.

Hayden-Wing, L. D. (1979). Elk use of mountain meadows in the Idaho Primitive Area. In *North American Elk: Ecology, Behavior and Management*, ed. M. S. Boyce & L. D. Hayden-Wing, pp. 40–6. Laramie: University of Wyoming.

Heller, J. (1986). Jackson: Values favored over commodities. *Jackson Hole Guide* (Jackson, Wyoming) 4 Feb 1986, pp. A1, A8–A9.

Hemming, J. E. (1971). *The Distribution and Movement Patterns of Caribou in Alaska*. Game Technical Bulletin No. 1, Juneau: Alaska Department of Fish and Game.

Hibler, C. P. (1981). Diseases. In *Mule and Black-Tailed Deer of North America*, ed. O. C. Wallmo, pp. 129–55. Lincoln: University of Nebraska Press.

Hieb, S. R. (1976). *Proceedings of the Elk–Logging–Roads Symposium*. Moscow: University of Idaho.

Hinds, T. E. (1985). Diseases. In *Aspen: Ecology and Management in the Western United States*, ed. N. V. DeByle & R. P. Winokur, pp. 87–106. US Department of Agriculture, Forest Service. General Technical Report RM–119. Fort Collins: Rocky Mountain Forest and Range Experiment Station.

Hobbs, N. T., Baker, D. L., Ellis, J. E. & Swift, D. M. (1979). Composition and quality of elk diets during winter and summer: A preliminary analysis. In *North American Elk: Ecology, Behavior and Management*, ed. M. S. Boyce & L. D. Hayden-Wing, pp. 47–53. Laramie: University of Wyoming.

Hobbs, N. T., Baker, D. L., Ellis, J. E., Smith, D. M. & Green, R. A. (1982). Energy- and nitrogen-based estimates of elk winter-range carrying capacity. *Journal of Wildlife Management*, **46**, 12–21.

Hobbs, N. T. & Swift, D. M. (1985). Estimates of habitat carrying capacity incorporating explicit nutritional constraints. *Journal of Wildlife Management*, **49**, 814–22.

Hocker, J. and Clark, S. (1981). *Jackson Hole: Protecting Public Values on Private Lands*. Jackson, Wyoming: Izaak Walton League of America.

Hornocker, M. G. (1970). An analysis of mountain lion predation on mule deer and elk in the Idaho Primitive Area. *Wildlife Monographs*, **21**.

Houston, D. B. (1968). *The Shiras Moose in Jackson Hole, Wyoming*. Technical Bulletin No. 1. Moose, Wyoming: Grand Teton Natural History Association.

Houston, D. B. (1978). Elk as winter–spring food for carnivores in northern Yellowstone National Park. *Journal of Applied Ecology*, **15**, 653–61.

Houston, D. B. (1979). The northern Yellowstone elk-winter distribution and management. In *North American Elk: Ecology, Behavior and Management*, ed. M. S. Boyce & L. D. Hayden-Wing, pp. 263–72. Laramie: University of Wyoming.

Houston, D. B. (1982). *The Northern Yellowstone Elk*. New York: MacMillan.

Hungerford, C. R. (1970). Response of Kaibab mule deer to management of summer range. *Journal of Wildlife Management*, **34**, 852–62.

Hutchinson, G. E. (1978). *An Introduction to Population Ecology*. New Haven, Connecticut: Yale University Press.

Inoue, M. & Kamifukumoto, H. (1984). Scenarios leading to chaos in a forced Lotka-Volterra model. *Progress of Theoretical Physics*, **71**, 930–7.

Irwin, L. L. & Peek, J. M. (1979). Relationships between road closures and elk behavior in Northern Idaho. In *North American Elk: Ecology, Behavior and Management*, ed. M. S. Boyce & L. D. Hayden-Wing, pp. 199–204. Laramie: University of Wyoming.

Irwin, L. L. & Peek, J. M. (1983a). Elk, *Cervus elaphus*, foraging related to forest management and succession in Idaho. *Canadian Field-Naturalist*, **97**, 443–7.

Irwin, L. L. & Peek, J. M. (1983b). Elk habitat use relative to forest succession in Idaho. *Journal of Wildlife Management*, **47**, 664–72.

Ise, J. (1961). *Our National Park Policy: A Critical History*. Baltimore: Johns Hopkins University Press.

Janson, R. G. (1966). Movements of transplanted Yellowstone elk in west-central Montana. *Proceedings of the Annual Conference of the Western Association of State Game and Fish Commissioners*, **46**, 107–9.

Jenkins, K. J. & Wright, R. G. (1988). Resource partitioning and competition among cervids in the Northern Rocky Mountains. *Journal of Applied Ecology*, **25**, 11–24.

Johnson, D. E. (1951). Biology of the elk calf, *Cervus canadensis nelsoni*. *Journal of Wildlife Management*, **15**, 396–410.

Johnson, R. (1976). Elk and logging activities. *Wyoming Wildlife*, **40**(11), 8–11.

Jones, D. A. (1985). Big game management from the US Forest Service perspective. In *Western Elk Management: A Symposium*, ed. G. W. Workman, pp. 69–73. Logan: Utah State University.

Jones, W. B. (1965). Response of major plant species to elk and cattle grazing in northwestern Wyoming. *Journal of Range Management*, **18**, 218–20.

Julander, O., Robinette, W. L. & Jones, D. A. (1961). Relation of summer range condition to mule deer herd productivity. *Journal of Wildlife Management*, **25**, 54–60.

Kay, C. (1985). Aspen reproduction in the Yellowstone Park – Jackson Hole area and its relationship to the natural regulation of ungulates. In *Western Elk Management: A Symposium*, ed. G. W. Workman, pp. 131–60. Logan: Utah State University.

Keiss, R. E. (1969). Comparison of eruption-wear patterns and cementum annuli as age criteria in elk. *Journal of Wildlife Management*, **33**, 175–80.

Kie, J. G. & White, M. (1985). Population dynamics of white-tailed deer (*Odocoileus virginianus*) on the Welder Wildlife Refuge, Texas. *Southwestern Naturalist*, **30**, 105–18.

Kistner, T. P., Greer, K. R., Worley, D. E. & Brunetti, O. A. (1982). Diseases and parasites. In *Elk of North America: Ecology and Management*, ed. J. W. Thomas & D. E. Toweill, pp. 181–217. Harrisburg, Pennsylvania: Stackpole Books.

Klein, D. R. (1965). Ecology of deer range in Alaska. *Ecological Monographs*, **35**, 259–84.

Klein, D. R. (1970). Food selection by North American deer and their response to over-utilization of preferred plant species. In *Animal Populations in Relation to Their Food Resources*, ed. A. Watson, pp. 25–46. Oxford: Blackwells.

Knight, R. R. (1966). The effectiveness of neckbands for marking elk. *Journal of Wildlife Management*, **30**, 845–6.

Knight, R. R. (1970). The Sun River elk herd. *Wildlife Monographs*, **23**, 1–66.

Krebill, R. G. (1972). *Mortality of Aspen on the Gros Ventre Elk Winter Range*. US Department of Agriculture Forest Service Research Paper INT–129, Ogden, Utah: Intermountain Forest and Range Experiment Station.

Kruskal, J. B. (1964a). Multidimensional scaling by optimizing goodness of fit to a nonmetric hypothesis. *Psychometrika*, **29**, 1–27.

Kruskal, J. B. (1964b). Nonmetric multidimensional scaling: A numerical method. *Psychometrika*, **29**, 115–29.

Kuck, L., Hompland, G. L. & Merrill, E. H. (1985). Elk calf response to simulated mine disturbance in southeast Idaho. *Journal of Wildlife Management*, **49**, 751–7.

Laycock, W. A. & Richardson, B. Z. (1975). Long-term effects of pocket gopher control on vegetation and soils of a subalpine grassland. *Journal of Range Management*, **28**, 458–62.

Leege, T. A. (1968). Prescribed burning for elk in northern Idaho. *Proceedings of the Annual Tall Timbers Fire Ecology Conference*, **8**, 235–53.

Leege, T. A. & Hickey, W. O. (1977). *Elk-Snow-Habitat Relationships in the Pete King Drainage, Idaho*. Wildlife Bulletin No. 6. Boise: Idaho Department of Fish and Game.

Leege, T. A. (1984). *Guidelines for Evaluating and Managing Summer Elk Habitat*. Wildlife Bulletin No. 11. Boise: Idaho Department of Fish and Game.

Leigh, E. G., Jr. (1970). Sex ratio and differential mortality between the sexes. *American Naturalist*, **104**, 205–10.

Leslie, D. M., Jr. (1983). *Nutritional ecology of cervids in old-growth forests in Olympic National Park, Washington*. PhD dissertation. Corvallis, Oregon: Oregon State University.

Long, B., Hinschberger, M., Roby, G. & Kimbal, J. (1980). *Gros Ventre Cooperative Elk Study. Final Report 1974–1979*. Jackson, Wyoming: Wyoming Game and Fish Department and the US Forest Service.

Lovaas, A. L. (1970). *People and the Gallatin Elk Herd*. Helena: Montana Game and Fish Department.

Lovaas, A. L., Egan, J. L. & Knight, R. R. (1966). Aerial counting of two Montana elk herds. *Journal of Wildlife Management*, **30**, 364–9.

Lyon, L. J. (1976). Elk use as related to characteristics of clearcuts in western Montana. In *Proceedings Elk-Logging-Roads Symposium*, ed. S. R. Hieb, pp. 69–72. Moscow: University of Idaho.

Lyon, L. J. (1979). *Influences of Logging and Weather on Elk Distribution in Western Montana*. US Department of Agriculture, Forest Service Research Paper, INT–236.

Lyon, L. J. & Ward, A. L. (1982). Elk and land management. In *Elk of North America: Ecology and Management*, ed. J. W. Thomas & D. E. Toweill, pp. 443–77. Harrisburg, Pennsylvania: Stackpole Books.

MacNab, J. (1985). Carrying capacity and related slippery shibboleths. *Wildlife Society Bulletin*, **13**, 403–10.

McMillan, J. F. (1953). Measures of association between moose and elk on feeding grounds. *Journal of Wildlife Management*, **17**, 162–6.

McNamara, K. (1979). Differences between quaking aspen genotypes in relation to browsing preference by elk. In *North American Elk: Ecology, Behavior and Management*, ed. M. S. Boyce & L. D. Hayden-Wing, pp. 83–8. Laramie: University of Wyoming.

Mackie, R. J. (1985). The elk-deer-livestock triangle. In *Western Elk Management: A Symposium*, ed. G. W. Workman, pp. 51–6. Logan: Utah State University.

Madsen, R. L. (1985). The history of supplemental feeding at the National Elk Refuge. In *Western Elk Management: A Symposium*, ed. G. W. Workman, pp. 45–9. Logan: Utah State University.

Marcum, C. L. (1976). Habitat selection and use during summer and fall months by a western Montana elk herd. In *Proceedings Elk-Logging-Roads Symposium*, ed. S. R. Hieb, pp. 91–6. Moscow: University of Idaho.

Marcum, C. L. (1979). Summer-fall food habits and forage preferences of a western Montana elk herd. In *North American Elk: Ecology, Behavior and Management*, ed. M. S. Boyce & L. D. Hayden-Wing, pp. 54–62. Laramie: University of Wyoming.

Marcum, C. L. & Scott, M. D. (1985). Influences of weather on elk use of spring-summer habitat. *Journal of Wildlife Management*, **49**, 73–6.

Martinka, C. J. (1965). *Population Status, Social Habits, Movements and Habitat Relationships of the Summer Resident Elk of Jackson Hole Valley, Wyoming*. MS thesis. Bozeman: Montana State University.

Martinka, C. J. (1969). Population ecology of summer resident elk in Jackson Hole, Wyoming. *Journal of Wildlife Management*, **33**, 465–81.

Martinka, C. J. (1978). Ungulate populations in relation to wilderness in Glacier National Park, Montana. *Transactions of the North American Wildlife and Natural Resources Conference*, **43**, 351–7.

Meagher, M. M. (1973). The bison of Yellowstone National Park. *National Park Service Scientific Monograph Series*, **1**, 1–161.

Mech, L. D. (1970). *The Wolf*. Garden City, New York: Natural History Press.

Mech, L. D. & Karns, P. D. (1977). *Role of the Wolf in a Deer Decline in the Superior National Forest*. Resource Paper NC–141. Washington, DC: US Department of Agriculture, Forest Service.

Melnykovych, A. (1985). Officials blame logging, new roads for elk decline near Jackson Hole. *Casper Star Tribune*, 4 May 1985, A4, A12.

Mendelssohn, R. (1976). Optimization problems associated with a Leslie matrix. *American Naturalist*, **110**, 339–49.

Menkens, G. E., Jr. & Boyce, M. S. (1989). Comments on the use of time-specific and cohort life tables. *Ecology*, (submitted).

Merrifield, J. & Gerking, S. (1982). *Analysis of the Long-Term Impacts and Benefits of Grand Teton National Park on the Economy of Teton County, Wyoming*. Laramie: University of Wyoming – National Park Service Research Center.

Merrifield, J. (1983). Using analog regions to assess the economic impact of federal land management policies. *Professional Geographer*, **35**, 298–302.

Mielke, J. L. (1957). *Aspen Leaf Blight in the Intermountain Region*. US Department of Agriculture, Forest Service, Research Note No. 42. Ogden, Utah: Intermountain Forest and Range Experiment Station.

Mohler, J. R. (1917). *Annual Report to the US Bureau of Animal Industry*. Washington, DC: US Department of Agriculture.

Morgantini, L. E. & Hudson, R. J. (1979). Human disturbance and habitat selection in elk. In *North American Elk: Ecology, Behavior and Management*, ed. M. S. Boyce & L. D. Hayden-Wing, pp. 132–9. Laramie: University of Wyoming.

Morgantini, L. E. & Hudson, R. J. (1985). Changes in diets of elk during a hunting season. *Journal of Range Management*, **38**, 77–9.

Morrison, J. A., Trainer, C. E. & Wright, P. L. (1959). Breeding season in elk as determined from known age embryos. *Journal of Wildlife Management*, **23**, 27–34.

Morton, J. K., Thorne, E. T. & Thomas, G. M. (1981). Brucellosis in elk. III. Serologic evaluation. *Journal of Wildlife Diseases*, **17**, 23–31.

Mould, E. D. & Robbins, C. T. (1981). Nitrogen metabolism in elk. *Journal of Wildlife Management*, **45**, 323–34.

Mueggler, W. F. (1972). Influence of competition on the response of bluebunch wheatgrass to clipping. *Journal of Range Management*, **25**, 88–92.

Mueggler, W. F. (1975). Rate and pattern of vigor recovery in Idaho fescue and bluebunch wheatgrass. *Journal of Range Management*, **28**, 198–204.

Murie, O. J. (1951a). Do you want hunting in our National Parks? *Natural History*, **60**, 464–7.

Murie, O. J. (1951b). Grand Teton National Park and its elk. *National Parks Magazine*, **25**(107), 119–20.

Murie, O. J. (1951c). *The Elk of North America*. Harrisburg, Pennsylvania: Stackpole Press. (reprinted by Teton Bookshop, 1979).

Murie, O. J. (1952). Elk shooting in Grand Teton again. *National Parks Magazine*, **26**(110), 118.

Murie, O. J. (1953). Teton's elk problem continues. *National Parks Magazine*, **27**(113), 56–67.

Murie, M. E. & Murie, O. (1966). *Wapiti Wilderness*. New York: Alfred A. Knopf.

Myers, N. (1979). *The Sinking Ark*. Oxford: Pergamon.

National Elk Refuge. (1966). *Annual Narrative Report*. Jackson, Wyoming: US Fish & Wildlife Service.

National Elk Refuge. (1984). *Annual Narrative Report*. Jackson, Wyoming: US Fish & Wildlife Service.

Neff, D. J. (1968). The pellet-group count technique for big-game trend, census, and distribution: A review. *Journal of Wildlife Management*, **32**, 597–614.

Neiland, K. A. (1970). Rangiferine brucellosis in Alaskan canids. *Journal of Wildlife Diseases*, **6**, 136–9.

Neiland, K. A. (1975). Further observations in rangiferine brucellosis in Alaskan carnivores. *Journal of Wildlife Diseases*, **11**, 45–52.

Nelson, J. R. (1982). Relationships of elk and other large herbivores. In *Elk of North America: Ecology and Management*, ed. J. W. Thomas & D. E. Toweill, pp. 415–41. Harrisburg, Pennsylvania: Stackpole Books.

Nelson, J. R. & Leege, T. A. (1982). Nutritional requirements and food habits. In *Elk of North America: Ecology and Management*, ed. J. W. Thomas & D. E. Toweill, pp. 323–67. Harrisburg, Pennsylvania: Stackpole Books.

Nelson, L. J. & Peek, J. M. (1982). Effect of survival and fecundity on rate of increase of elk. *Journal of Wildlife Management*, **46**, 535–40.

Nesbitt, W. H. & Parker, J. S. (1977). *North American Big Game*. Washington, D.C.: Boone and Crockett Club and National Rifle Association.

Nesbitt, W. H. & Wright, P. L. (1981). *Records of North American Big Game*. Alexandria, Virginia: Boone and Crockett Club.

Oakley, C. A. (1975). Elk distribution in relation to a deferred rotation grazing system. *Journal of Range Management*, **28**, 274 [abstract].

Oldemeyer, J. L., Barmore, W. J. & Gilbert, D. L. (1971). Winter ecology of bighorn sheep in Yellowstone National Park. *Journal of Wildlife Management*, **35**, 257–69.

Olmsted, C. E. (1979). The ecology of aspen with reference to utilization by large herbivores in Rocky Mountain National Park. In *North American Elk: Ecology, Behavior and Management*, ed. M. S. Boyce & L. D. Hayden-Wing, pp. 89–97. Laramie: University of Wyoming.

Ozoga, J. J. & Verme, L. J. (1982). Physical and reproductive characteristics of a supplementally-fed white-tailed deer herd. *Journal of Wildlife Management*, **46**, 281–301.

Ozoga, J. J. (1987). Maximum fecundity in supplementally-fed Northern Michigan white-tailed deer. *Journal of Mammalogy*, **68**, 878–9.

Palmer, M. A., Ostry, M. E. & Schipper, A. L., Jr. (1980). *How to Identify and Control* Marssonina *Leaf Spot of Poplars*. US Department of Agriculture, Forest Servcice. Washington, DC: US Government Printing Office.

Paloheimo, J. E. & Fraser, D. (1981). Estimation of harvest rate and vulnerability from age and sex data. *Journal of Wildlife Management*, **45**, 948–58.

Patton, D. R. & Jones, J. R. (1977). *Managing Aspen for Wildlife in the Southwest*. US Department of Agriculture, Forest Service. General Technical Report RM–37. Fort Collins, Colorado: Rocky Mountain Forest and Range Experiment Station.

Peek, J. M. & Lovaas, A. L. (1968). Differential distribution of elk by sex and age on the Gallatin Winter Range, Montana. *Journal of Wildlife Management*, **32**, 553–7.

Peek, J. M., Scott, M. D., Nelson, L. J., Pierce, D. J. & Irwin, L. L. (1982). Role of cover in habitat management for big game in northwestern United States. *Transactions of the North American Wildlife and Natural Resources Conference*, **47**, 363–73.

Peek, J. M. (1985). On counting elk. *Bugle*, **2**(1), 46–7.

Peek, J. M. & Scott, M. D. (1985). Elk and cover. In *Western Elk Management: A Symposium*, ed. G. W. Workman, pp. 75–82. Logan: Utah State University.

Peek, J. M. (1986). *A Review of Wildlife Management*. Englewood Cliffs, New Jersey: Prentice Hall.

Pellew, R. A. (1983). Modeling and the systems approach to management problems: The *Acacia*/elephant problem in the Serengeti. In *Management of Large Mammals in African Conservation Areas*, ed. R. N. Owen-Smith, pp. 93–114. Pretoria, South Africa: Haum.

Pengelly, W. L. (1963). Thunder on the Yellowstone. *Naturalist*, **14**, 18–25.

Perala, D. A. (1984). How endemic injuries affect early growth of aspen suckers. *Canadian Journal of Forest Research*, **14**, 755–62.

Phillips, C. & Ferguson, S. (1977). *Hunting and Fishing Expenditure Values and Participation Preferences in Wyoming, 1975*. Water Resources Research Institute. Laramie: University of Wyoming.

Picton, H. D. (1960). Migration patterns of the Sun River Elk Herd, Montana. *Journal of Wildlife Management*, **24**, 279–90.

Picton, H. D. (1984). Climate and the prediction of reproduction of three ungulate species. *Journal of Applied Ecology*, **21**, 869–79.

Pielou, E. C. (1975). *Ecological Diversity*. New York: Wiley.

Pimm, S. L. (1987). Determining the effects of introduced species. *Trends in Ecology and Evolution*, **2**, 106–8.

Pinon, J. & Poissonnier, M. (1975). Epidemiological study of *Marssonina brunnea* (Ell. & Ev.) P. Magn. *European Journal of Forest Pathology*, **5**, 97–111.

Pojar, T. M. (1981). A management perspective of population modeling. In *Dynamics of Large Mammal Populations*, ed. C. W. Fowler & T. D. Smith, pp. 241–61. New York: John Wiley.

Potter, D. R. (1982). Recreational uses of elk. In *Elk of North America: Ecology and Management*, ed. J. W. Thomas & D. E. Toweill, pp. 509–59. Harrisburg, Pennsylvania: Stackpole Books.

Preble, E. A. (1911). *Report on Condition of Elk in Jackson Hole, Wyoming in 1911*. Bulletin 40. Washington, DC: US Department of Agriculture, Bureau of Biological Survey.

Premer, G. E., Lewis, E. P., Fletcher, R. R. & Taylor, D. L. (1979). *Recreation and Tourism in the Teton County Economy*, Wyoming Agricultural Extension Service, Bulletin No. 704. Laramie: University of Wyoming.

Prevedel, D. A., Robison, M. H. & Iverson, D. C. (1985). *Preliminary Impact Analysis of the Dubois, Wyoming Community Relative to National Forest Management Decisions*. Intermountain Region. Ogden, Utah: US Department of Agriculture, Forest Service.

Price, M. A. & White, R. G. (1985). Growth and development. In *Bioenergetics of Wild Ruminants*, ed. R. J. Hudson & R. G. White, pp. 183–213. Boca Raton, Florida: CRC Press.

Price, P. W., Slobodchikoff, C. N. & Gaud, W. S. (1984). *A New Ecology*. New York: Wiley.

Prothero, W. L., Spillett, J. J. & Balph, D. F. (1979). Rutting behavior of yearling and mature bull elk: Some implications for open bull hunting. In *North American Elk: Ecology, Behavior and Management*, ed. M. S. Boyce & L. D. Hayden-Wing, pp. 160–5. Laramie: University of Wyoming.

Ransom, A. B. (1967). Reproductive biology of white-tailed deer in Manitoba. *Journal of Wildlife Management*, **31**, 114–23.

Rausch, R. A. & Hinman, R. A. (1977). Wolf management in Alaska – an exercise in futility. In *Proceedings 1975 Predator Symposium*, ed. R. L. Phillips & C. Jonkel, pp. 147–56. Missoula, Montana: Forest and Conservation Experiment Station.

Reese, J. B., Mohr, F. R., Dean, R. E. & Klabunde, T. (1975). *Teton Wilderness Fire Management Plan*. Jackson, Wyoming: US Department of Agriculture, Forest Service, Bridger–Teton National Forest.

Reimers, E. (1972). Growth in domestic and wild reindeer in Norway. *Journal of Wildlife Management*, **36**, 612–19.

Repanshek, K. J. (1986). Greater Yellowstone Coalition warned of oil, gas leasing threats. *Laramie Daily Boomerang* (Laramie, Wyoming), **106**, 11.

Reynolds, H. G. (1966). *Use of Openings in Spruce-Fir Forests of Arizona by Elk, Deer and Cattle*. US Department of Agriculture, Forest Service Research Note RM–66. Fort Collins, Colorado: Rocky Mountain Forest and Range Experiment Station.

Righter, R. W. (1982). *Crucible for Conservation: The Creation of Grand Teton National Park*. Denver: Colorado Assoc. Univ. Press.

Rippe, D. J. & Rayburn, R. L. (1981). *Land Use and Big Game Population Trends in Wyoming*. US Department of Interior, Fish and Wildlife Service. Biological Services Program W/CRAM – 81/W22. Fort Collins, Colorado: Western Energy and Land Use Team.

Roach, M. E. (1950). Estimating perennial grass utilization on semi-desert cattle ranges by percentage of ungrazed plants. *Journal of Range Management*, **3**, 182–5.

Robbins, R. L. & Wilbrecht, J. (1979). Supplemental feeding of elk wintering on the National Elk Refuge. In *North American Elk: Ecology, Behavior and Management*, ed. M. S. Boyce & L. D. Hayden-Wing, pp. 255–8. Laramie: University of Wyoming.

Robbins, R. L., Redfearn, D. E. & Stone, C. P. (1982). Refuges and elk management. In *Elk of North America: Ecology and Management*, ed. J. W. Thomas & D. E. Toweill, pp. 479–507. Harrisburg, PA: Stackpole Books.

Robel, R. J. (1960). Determining elk movements through periodic aerial counts. *Journal of Wildlife Management*, **24**, 103–4.

Robinson, W. L. & Bolen, E. G. (1984). *Wildlife Ecology and Management*. London: Collier Macmillan Publishers.

Robson, D. S. (1960). An unbiased sampling and estimation procedure for creel census of fishermen. *Biometrics*, **16**, 261–77.

Roelle, J. E. & Auble, G. T. (1983). Resource development and management in Jackson Hole, Wyoming. In *Developments in Environmental Modelling, 5: Analysis of Ecological Systems: State of the Art in Ecological Modelling*, ed. W. K. Laurenroth, G. V. Skogeroboe & M. Flug, pp. 303–11. New York: Elsevier.

Romme, W. H. & Knight, D. H. (1981). Fire frequency and subalpine forest succession along a topographic gradient in Wyoming. *Ecology*, **62**, 319–26.

Romme, W. H. (1982). Fire and landscape diversity in subalpine forests of Yellowstone National Park. *Ecological Monographs*, **52**, 199–221.

Rudd, W. J., Ward, A. L. & Irwin, L. L. (1983). Do split hunting seasons influence elk migrations from Yellowstone National Park? *Wildlife Society Bulletin*, **11**, 328–31.

Rush, W. M. (1932). Bang's disease in Yellowstone National Park buffalo and elk herds. *Journal of Mammalogy*, **13**, 371–2.

Samuel, M. D. & Pollock, K. H. (1981). Correction of visibility bias in aerial surveys where animals occur in groups. *Journal of Wildlife Management*, **45**, 993–7.

Samuel, M. D., Garton, E. O., Schlegel, M. W. & Carson, R. G. (1987). Visibility bias during aerial surveys of elk in north central Idaho. *Journal of Wildlife Management*, **51**, 622–30.

Sauer, J. R. & Boyce, M. S. (1979a). Elk-cattle interactions in calving areas. *University of Wyoming – National Park Service Research Center Annual Report*, **3**, 51–3.

Sauer J. R. & Boyce, M. S. (1979b). Time-series analysis of the National Elk Refuge census.

In *North American Elk: Ecology, Behavior and Management*, ed. M. S. Boyce & L. D. Hayden-Wing, pp. 9–12. Laramie: University of Wyoming.

Sauer, J. R. (1980a). Elk-cattle interactions on elk calving areas. *Journal of the Colorado-Wyoming Academy of Science*, **12**(1), 7 [abstract].

Sauer, J. R. (1980b). *The Population Ecology of Wapiti in Northwestern Wyoming*. MS Thesis. Laramie: University of Wyoming.

Sauer, J. R. & Boyce, M. S. (1983). Density-dependence and survival of elk in Northwestern Wyoming. *Journal of Wildlife Management*, **47**, 31–7.

Schaffer, W. M. (1982). Ecological abstraction: the consequences of reduced dimensionality in ecological models. *Ecological Monographs*, **51**, 383–401.

Schaffer, W. M. (1985). Can nonlinear dynamics elucidate mechanisms in ecology and epidemiology? *IMA Journal of Mathematics Applied in Medicine & Biology*, **2**, 221–52.

Schlegel, M. (1976). Factors affecting calf elk survival in north central Idaho: A progress report. *Proceedings of the Annual Conference of the Western Association of State Game and Fish Commissioners*, **56**, 342–55.

Schreuder, H. T., Tyre, G. L. & James, G. A. (1975). Instant- and interval-count sampling: two new techniques for estimating recreation use. *Forest Science*, **21**, 40–4.

Scott, K. W. & Wilson, M. (1984). Dead Indian Creek local fauna. *Wyoming Archaeologist*, **27**, 51–62.

Seastedt, T. R. (1985). Maximization of primary and secondary productivity by grazers. *American Naturalist*, **126**, 559–64.

Seber, G. A. F. (1970). The effects of trap response on tag-recapture estimates. *Biometrics*, **26**, 13–22.

Seber, G. A. F. (1973). *The Estimation of Animal Abundance and Related Parameters*. London: Griffen.

Severinghaus, C. W. & Moen, A. N. (1983). Prediction of weight and reproductive rates of a white-tailed deer population from records of antler beam diameter among yearling males. *New York Fish & Game Journal*, **30**, 30–8.

Shaw, R. J. (1976). *Field Guide to the Vascular Plants of Grand Teton National Park and Teton County, Wyoming*. Logan: Utah State University Press.

Sheldon, C. (1927). *The Conservation of the Elk of Jackson Hole, Wyoming*. Washington, DC: Report to the President's Committee on Outdoor Recreation and the Governor of Wyoming.

Shoesmith, M. W. (1979). Seasonal movements and social behavior of elk on Mirror Plateau, Yellowstone National Park. In *North American Elk: Ecology, Behavior and Management*, ed. M. S. Boyce & L. D. Hayden-Wing, pp. 166–76. Laramie: University of Wyoming.

Sinclair, A. R. E. (1983). Management of conservation areas as ecological baseline controls. In *Management of Large Mammals in African Conservation Areas*, ed. R. N. Owen-Smith, pp. 13–22. Pretoria, South Africa: Haum.

Sinclair, A. R. E. (1985). Does interspecific competition or predation shape the African ungulate community? *Journal of Animal Ecology*, **54**, 899–918.

Singer, F. J. (1979). Habitat partitioning and wildfire relationships of cervids in Glacier National Park, Montana. *Journal of Wildlife Management*, **43**, 437–44.

Skinner, S. (1985). Cry wolf. *Wyoming Wildlife*, **49**(10), 6–11.

Skogland, T. (1986). Sex ratio variation in relation to maternal condition and parental investment in wild reindeer *Rangifer t. tarandus*. *Oikos*, **46**, 417–19.

Skovlin, J. M., Edgerton, P. J. & Harris, R. W. (1968). The influence of cattle management on deer and elk. *Transactions of the North American Wildlife Conference*, **33**, 169–81.

Skovlin, J. M. (1982). Habitat requirements and evaluations. In *Elk of North America:*

Ecology and Management, ed. J. W. Thomas & D. E. Toweill, pp. 369–413. Harrisburg, Pennsylvania: Stackpole Books.

Slade, N. A. (1977). Statistical detection of density dependence from a series of sequential censuses. *Ecology*, **58**, 1094–1102.

Smith, B. L. & Robbins, R. L. (1984). *Pelleted alfalfa hay as supplemental winter feed for elk at the National Elk Refuge*. Jackson, Wyoming: US Department of Interior, Fish and Wildlife Service, National Elk Refuge.

Smith, B. L. (1985). Scabies and elk mortalities on the National Elk Refuge, Wyoming. *Western States and Provinces Elk Workshop Proceedings*, **1984**, 180–94.

Smith, D. R. & Wilbert, D. E. (1958). *Game/livestock Competition in the Jackson Hole Region of Wyoming*. Unpublished Report. Laramie: Wyoming Agricultural Experiment Station.

Smith, D. R. (1961). *Competition Between Cattle and Game on Elk Winter Range*. Agricultural Experiment Station Bulletin no. 377. Laramie: University of Wyoming.

Sokal, R. R. & Rohlf, F. J. (1981). *Biometry*, 2nd edn. San Francisco: W. H. Freeman and Company.

Sorg, C. F. & Loomis, J. (1985). An introduction to wildlife valuation techniques. *Wildlife Society Bulletin*, **13**, 38–46.

Stelfox, J. G. (1976). Range ecology of Rocky Mountain bighorn sheep. *Canadian Wildlife Service Report Series*, **39**, 1–50.

Stevens, D. R. (1966). Range relationships of elk and livestock, Crow Creek Drainage, Montana. *Journal of Wildlife Management*, **30**, 349–63.

Stevens, D. R. (1974). Rocky Mountain elk-Shiras moose range relations. *Canadian Naturalist*, **101**, 505–16.

Stoddart, L. A. & Smith, A. D. (1943). *Range management*. New York: McGraw Hill.

Straley, J. H. (1968). Population analysis of ear-tagged elk. *Proceedings of the Western Association of State Game and Fish Commissioners*, **48**, 152–60.

Straley, J. H., Roby, G. & Johnson, B. (1983). *Annual Big Game Unit Reports. District I.* Cheyenne: Wyoming Game and Fish Department.

Straley, J. H., Roby, G. & Johnson, B. (1984). *Annual Big Game Unit Reports. District I.* Cheyenne: Wyoming Game and Fish Department.

Sweeney, J. M. (1976). *Elk Movements and Calving as Related to Snow Cover*. PhD dissertation. Fort Collins: Colorado State University.

Sweeney, J. M. & Steinhoff, H. W. (1976). Elk movements and calving as related to snow cover. In *Ecological Impacts of Snowpack Augmentation in the San Juan Mountains, Colorado*, ed. H. W. Steinhoff & J. D. Ives, pp. 415–36. Fort Collins: Colorado State University Publications.

Swift, D. M. (1983). A simulation model of energy and nitrogen balance for free-ranging ruminants. *Journal of Wildlife Management*, **47**, 620–45.

Taber, R. D., Raedeke, K. & McCaughran, D. A. (1982). Population characteristics. In *Elk of North America: Ecology and Management*, ed. J. W. Thomas & D. E. Toweill, pp. 279–98. Harrisburg, Pennsylvania: Stackpole Books.

Taylor, D. T. & Bradley, E. B. (1982). *Recreation and Tourism in the Jackson Hole Area*. Wyoming Agricultural Extension Service, Bulletin No. 787. Laramie: University of Wyoming.

Taylor, D. T., Bradley, E. B. & Martin, M. M. (1982). *The Outfitting Industry in Teton County: Its Clientele and Economic Importance*. Wyoming Agricultural Extension Service, Bulletin No. 793. Laramie: University of Wyoming.

Teton County Board of Commissioners (1980). *Teton County Comprehensive Plan and Implementation Program*. Jackson, Wyoming: Teton County Commissioners.

Thomas, E., Crowe, D. & Kruckenberg, L. L. (1984). Managing Teton elk. *Wyoming Wildlife*, **48**(5), 18–21.

Thomas, J. W. 1979. *Wildlife Habitats in Managed Forests. The Blue Ridge Mountains of Oregon and Washington*. US Department of Agriculture, Forest Service Agriculture Handbook No. 553. Washington, DC: US Government Printing Office.

Thomas, J. W. & Sirmon, J. M. (1985). Keys to the future of elk and elk hunting. *Bugle*, **2**(3), 22–5.

Thorne, E. T. & Morton, J. K. (1975). *The Incidence and Importance of Brucellosis in Elk in Northwestern Wyoming*. P-R Job Progress Report, Project FW–3–R–21, Work Plan No. 1, Job No. 8W. Cheyenne: Wyoming Game and Fish Department.

Thorne, E. T. & Butler, G. (1976). *Comparison of Pelleted, Cubed, and Baled Alfalfa Hay as Winter Feed for Elk*. Wildlife Technical Report No. 6. Cheyenne: Wyoming Game and Fish Department.

Thorne, E. T., Dean, R. E. & Hepworth, W. G. (1976). Nutrition during gestation in relation to successful reproduction in elk. *Journal of Wildlife Management*, **40**, 330–5.

Thorne, E. T., Morton, J. K. & Thomas, G. M. (1978). Brucellosis in elk. I. Serologic and bacteriologic survey in Wyoming. *Journal of Wildlife Diseases*, **14**, 74–81.

Thorne, E. T., Morton, J. K., Blunt, F. M. & Dawson, H. A. (1978). Brucellosis in elk. II. Clinical effects and means of transmission as determined through artificial insemination. *Journal of Wildlife Diseases*, **14**, 280–91.

Thorne, E. T., Morton, J. K. & Ray, W. C. (1979). Brucellosis, its affect and impact on elk in western Wyoming. In *North American Elk: Ecology, Behavior and Management*, ed. M. S. Boyce & L. D. Hayden-Wing, pp. 212–20. Laramie: University of Wyoming.

Thorne, E. T. (1981). *Importance of Nutrition During Late Gestation to Reproduction in Elk*. Federal Aid Report, Project FW–3–R–26, Work Plan 2, Job 10W. Cheyenne: Wyoming Game & Fish Department.

Thorne, E. T., Walthall, T. J. & Dawson, H. A. (1981). Vaccination of elk with strain 19 *Brucella abortus*. *Proceedings US Animal Health Association*, **85**, 359–74.

Thorne, E. T., Kingston, N., Jolley, W. R. & Bergstrom, R. C. (1982). *Diseases of Wildlife in Wyoming*. Cheyenne: Wyoming Game and Fish Department.

Thorne, E. T. (1984). Elk of North America: Ecology and Management (book review). *Journal of Wildlife Management*, **48**, 663–4.

Thorne, E. T. & Anderson, S. L. (1984). Immune response of elk vaccinated with a reduced dose of strain 19 Brucella vaccine. In *Game and Fish Research*, pp. 37–47. Cheyenne: Wyoming Game and Fish Department.

Thuermer, A. M., Jr. (1985). Amoco proposes well. *Jackson Hole News*, 30 Oct 1985, 3.

Thuermer, A. M., Jr. (1986a). Feds seize McReynolds-Thompson land in Grand Teton Park. *Jackson Hole News*, 15 Jan 1986, 3.

Thuermer, A. M., Jr. (1986b). Tixier calls Mosquito Creek 'made to order' drill site. *Jackson Hole News*, 21 May 1986, 5.

Trivers, R. L. & Willard, D. E. (1973). Natural selection of parental ability to vary the sex ratio of offspring. *Science*, **179**, 90–2.

US Fish & Wildlife Service. (1984). *Agency Review Draft Revised Northern Rocky Mountain Wolf Recovery Plan*. Prepared by the Northern Rocky Mountain Wolf Recovery Team. Denver, Colorado, USA: US Fish & Wildlife Service.

Verme, L. J. (1963). Effect of nutrition on growth of white-tailed deer fawns. *Transactions of the North American Wildlife and Natural Resources Conference*, **28**, 431–43.

Verme, L. J. (1965). Reproduction studies on penned white-tailed deer. *Journal of Wildlife Management*, **29**, 74–9.

Verme, L. J. (1967). Influence of experimental diets on white-tailed deer reproduction.

Transactions of the North American Wildlife and Natural Resources Conference, **32**, 405–20.

Verme, L. J. (1983). Sex ratio variation in *Odocoileus*: A critical review. *Journal of Wildlife Management*, **47**, 573–82.

Weaver, J. L. (1979a). Influence of elk carrion upon coyote populations in Jackson Hole, Wyoming. In *North American Elk: Ecology, Behavior and Management*, ed. M. S. Boyce & L. D. Hayden-Wing, pp. 152–7. Laramie: University of Wyoming.

Weaver, J. L. (1979b). Wolf predation upon elk in the Rocky Mountain parks of North America: A review. In *North American Elk: Ecology, Behavior and Management*, ed. M. S. Boyce & L. D. Hayden-Wing, pp. 29–33. Laramie: University of Wyoming.

Weinstein, J. (1979). The condition and trend of aspen along Pacific Creek in Grand Teton National Park. In *North American Elk: Ecology, Behavior and Management*, ed. M. S. Boyce & L. D. Hayden-Wing, pp. 79–82. Laramie: University of Wyoming.

Wells, M. C. (1979). *Wildlife in Jackson Hole: Private Lands as Critical Habitat*. US Department of Agriculture, Forest Service. Jackson, Wyoming: Bridger–Teton National Forest.

Westra, R. & Hudson, R. J. (1981). Digestive function of wapiti calves. *Journal of Wildlife Management*, **45**, 148–55.

White, G. C., Anderson, D. R., Burnham, K. P. & Otis, D. L. (1982). *Capture-recapture and Removal Methods for Sampling Closed Populations*. LA–8787–NERP. Los Alamos, New Mexico: Los Alamos National Laboratory.

Whitehead, G. K. (1982). *Hunting and Stalking Deer Throughout the World*. London: B. T. Batsford, Ltd.

Wickstrom, M. L., Robbins, C. T., Hanley, T. A., Spalinger, D. E. & Parish, S. M. (1984). Food intake and foraging energetics of elk and mule deer. *Journal of Wildlife Management*, **48**, 1285–1301.

Wilbert, D. E. (1959). *Range Studies on Elk Winter and Spring Ranges*. Federal Aid Report W–67–R–1. Job 2. Cheyenne: Wyoming Game and Fish Department.

Wilbrecht, J. & Robbins, R. L. (1979). History of the National Elk Refuge. In *North American Elk: Ecology, Behavior and Management*, ed. M. S. Boyce & L. D. Hayden-Wing, pp. 248–55. Laramie: University of Wyoming.

Wilbrecht, J. (1983). The National Elk Refuge: History and management including use of prescribed fire. In *Fire: Its Field Effects, Proceedings of a Symposium*, J. E. Lotan, pp. 103–13. Missoula, Montana: Intermountain Fire Council; and Pierre, South Dakota: Rocky Mountain Fire Council.

Wilson, S. S. (1958). *Seasonal Migration and Distribution of the Jackson Hole Elk Herd*. Federal Aid Report, W–27–R–11, Work Plan No. 8, Job No. 1. Cheyenne: Wyoming Game and Fish Commission.

Winn, D. S. (1976). *Terrestrial vertebrate fauna and selected coniferous habitat types on the north slope of the Uinta Mountains*. Wasatch National Forest Special Report. Salt Lake City, Utah: US Department of Agriculture, Forest Service.

Wolfe, M. L. (1975). Mortality patterns in the Isle Royale moose population. *American Midland Naturalist*, **97**, 267–79.

Wood, P. (1984). The elk hunt goes on at Grand Teton. *National Parks*, **58**(9–10), 29–31.

Wood, R. P. & Yorgason, I. J. (1974). *Elk Migration Study. Jackson Hole Elk Herd*. Moose, Wyoming: Grand Teton National Park; Cheyenne: Wyoming Game and Fish Department.

Wood, R. P. & Roby, G. (1975). *Elk Migration Study. Jackson Hole Elk Herd*. Moose, Wyoming: Grand Teton National Park; Cheyenne: Wyoming Game and Fish Department.

Wyoming Game and Fish Commission. (1984). *Annual Report*. Cheyenne: Wyoming Game and Fish Department.

Wyoming Game and Fish Department. (1959). *Annual Report of Big Game Harvest*. Cheyenne: Wyoming Game and Fish Department.

Wyoming Game and Fish Department. (1983). *A Strategic Plan for the Comprehensive Management of Wildlife in Wyoming, 1984–1989, Volume III*. Cheyenne: Wyoming Game and Fish Department.

Yorgason, I. J. (1963). *Range Studies on Elk Winter and Summer Ranges*. Federal Aid Report W–66–R–5. Job 3. Cheyenne: Wyoming Game and Fish Department.

Yorgason, I. J. & Cole, G. F. (1963). *Elk Migration Study. Jackson Hole Elk Herd*. Moose, Wyoming: Grand Teton National Park; Cheyenne: Wyoming Game and Fish Department.

Yorgason, I. J. (1966). The northern Jackson Hole elk herd. *Wyoming Wildlife*, **30**(3), 28–34.

Yorgason, I. J. & Cole, G. F. (1967). *Elk Migration Study. Jackson Hole Elk Herd*. Moose, Wyoming: Grand Teton National Park; Cheyenne: Wyoming Game and Fish Department.

Yorgason, I. J. (1969). *Technical Committee Report*. Proceedings Jackson Hole Cooperative Elk Studies Advisory Council Meeting, 6 March 1969, pp. 1–12. Jackson, Wyoming: Jackson Hole Cooperative Elk Studies Group.

NB: Most reports published by the Wyoming Game and Fish Department are available from Game Division, Wyoming Game and Fish Department, Cheyenne, WY 82003 USA. Some early reports on the Jackson elk herd had limited distribution; a reasonably complete collection is maintained by the Library, Grand Teton National Park, Moose, WY 83012 USA.

AUTHOR INDEX

SUBJECT INDEX